Other titles in this series

Island Ecology	M. Gorman
Insect Herbivory	I.D. Hodkinson and M.K. Hughes
Modelling	John N.R. Jeffers
Vegetation Dynamics	John Miles
Animal Population Dynamics	R. Moss, A. Watson and J.Ollason

Outline

Editors

George M. Dunnet
Regius Professor of Natural Hist(
University of Aberdeen

Charles H. Gimingham
Professor of Botany,
University of Aberdeen

Editors' Foreword

Both in its theoretical and applied aspects, ecology is developing rapidly. This is partly because it offers a relatively new and fresh approach to biological enquiry; it also stems from the revolution in public attitudes towards the quality of the human environment and the conservation of nature. There are today more professional ecologists than ever before, and the number of students seeking courses in ecology remains high. In schools as well as universities the teaching of ecology is now widely accepted as an essential component of biological education, but it is only within the past quarter of a century that this has come about. In the same period, the journals devoted to publication of ecological research have expanded in number and size, and books on aspects of ecology appear in ever-increasing numbers.

These are indications of a healthy and vigorous condition, which is satisfactory not only in regard to the progress of biological science but also because of the vital importance of ecological understanding to the well-being of man. However, such rapid advances bring their problems. The subject develops so rapidly in scope, depth and relevance that text-books, or parts of them, soon become out-of-date or inappropriate for particular courses. The very width of the front across which the ecological approach is being applied to biological and environmental questions introduces difficulties: every teacher handles his subject in a different way and no two courses are identical in content.

This diversity, though stimulating and profitable, has the effect that no single text-book is likely to satisfy fully the needs of the student attending a course in ecology. Very often extracts from a wide range of books must be consulted, and while this may do no harm it is time-consuming and expensive. The present series has been designed to offer quite a large number of relatively small booklets, each on a restricted topic of fundamental importance which is likely to constitute a self-contained component of more comprehensive courses. A selection can then be made, at reasonable cost, of texts appropriate to particular courses or the interests of the reader. Each is written by an acknowledged expert in the subject and is intended to offer an up-to-date, concise summary which will be of value to those engaged in teaching, research or applied ecology as well as to students.

Studies in Ecology

Plant–Atmosphere Relationships

J. GRACE

Department of Forestry and Natural Resources
University of Edinburgh

SPRINGER-SCIENCE+
BUSINESS MEDIA, B.V.

© 1983 J. Grace

ISBN 978-0-412-23180-3   ISBN 978-94-011-8048-1 (eBook)
DOI 10.1007/978-94-011-8048-1

British Library Cataloguing in Publication Data

Grace, John
    Plant–atmosphere relationships.—(Outline studies
    in ecology)
    1. Botany—Ecology
    I. Title   II. Series
    581.5   QK901
    ISBN 978-0-412-23180-3

Library of Congress Cataloging in Publication Data

Grace, John, 1945–
    Plant–atmosphere relationships

    (Outline studies in ecology)
    Bibliography: p.
    Includes index.
    1. Plant–atmosphere relationships. I. Title.
II. Series.
QK754.4.G7   1983   581.1′0427   82-19717
ISBN 978-0-412-23180-3

Contents

Preface

In this small book I have tried to confine myself to the absolute necessities in a field which requires a knowledge of both biology and physics. It is meant as a primer for biological undergraduates. I hope it will lead some of them to further, more advanced, study.

It has not been easy to present the subject in so few pages, and I am aware of many omissions. I hope readers will agree that it is best to concentrate on a small number of topics, which together constitute an essay on plant–atmosphere relationships. Advanced students will be able to take the subject further if they look up some of the references. Text books that I particularly recommend are those by Monteith [38] and Campbell [100]. If the reader intends to carry out research investigations he should also consult Fritschen and Lloyd [105] for an introduction to instrumentation in environmental biophysics.

The subject of plant–atmosphere relationships has developed considerably in recent years and is now at a most interesting stage. Most of the earlier micrometeorological work was concerned with agricultural crops, in which interest centred on water and carbon dioxide exchanges and their relationship to yield. However, the ecologist's horizon is broader than this, extending beyond the monoculture to problems of adaptation and survival of plants in more complex situations. In the last few years much more attention has been given to the biophysical analysis of wild systems (see for example recent issues of the journal *Oecologia Plantarum*). With the advent of cheap data logging and microprocessor-controlled data-acquisition, we should soon obtain important insights into the intricacies of plant–atmosphere relations in mixed communities of plants.

I would like to dedicate this book to the students in the Department of Forestry and Natural Resources at Edinburgh University, without whom the need to write a little book like this would not have been apparent. I would also like to acknowledge the stimulus that has come from postgraduate students and colleagues, and thank all those who gave up their time to comment on the original manuscript. Thanks are also due to Kath Higham who typed the manuscript with painstaking care.

John Grace
June 1982

1 Conditions for life

Be not afraid, the isle is full of noises,
Sounds, and sweet airs, that give delight and hurt not.
Shakespeare (part of Caliban's speech, *The Tempest*)

1.1 Radiant energy

Solar radiation is the primary source of energy on Earth, and life depends on it. We are well aware of the flux of radiant energy from the sun as about half of it is in the part of the spectrum sensed by our eyes. We are generally much less aware of the emission of longwave radiant energy by all surfaces around us. The first part of this chapter briefly introduces these two components of radiation which play such an important role in the heat balance of vegetation, to be considered in later chapters.

1.1.1 Solar radiation

The sun is an incandescent body, powered by a complex series of nuclear reactions, with a surface temperature of about 5800 K. There are small variations ($\pm 1\%$) in the output of the sun associated with various phenomena on the solar surface, especially sunspots, which vary in a cycle of about 11 years. We know little about long term changes in solar output, though it is acknowledged that they may have been large and could have had a great effect on conditions on Earth. Also, there are some changes in the Earth's orbit which varies from a near-circle to an ellipse, in a cycle of about 100 000 years. Moreover the Earth's axis wobbles in a cycle of 21 000 years. Hence, there are definite long term cycles in the rate at which solar energy reaches the Earth, in addition to global and seasonal variation. Above the atmosphere the solar flux density is at present about 1355 W m^{-2}, a quantity known as the solar constant even though it varies somewhat.

The solar spectrum contains energy within the ultraviolet, visible and near infra-red wavebands (Fig. 1.1). The radiation is considerably modified by passage through the atmosphere. An overall attenuation occurs, and there is a selective absorption of certain parts of the spectrum by atmospheric gases. As the path length of the solar beam through the atmosphere varies with solar angle, the spectrum alters with time of day, especially near sunrise and sunset.

Particles in the atmosphere scatter the solar beam and give rise to sky-light or diffuse radiation. Particles that are very small in relation to the wavelength of light, such as molecules, exhibit Rayleigh scattering in which the shorter wavelengths are scattered most, giving rise to blue sky. Larger particles varying in radius from a tenth of a wavelength to 25 wavelengths exhibit Mie scattering, in which, for the larger particles, all wavelengths are scattered equally, causing the sky to appear white as it

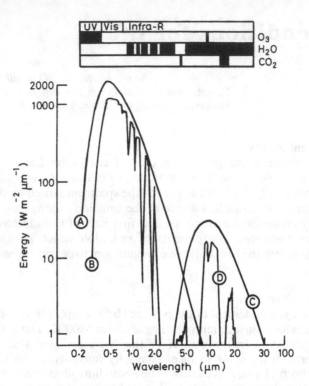

Fig. 1.1 Spectral energy distribution of radiation in the ultraviolet (UV), visible (Vis) and infra-red (Infra-R) parts of the spectrum: A, solar flux outside the Earth's atmosphere; B, solar flux at ground level after attenuation by gases in the atmosphere; C, typical longwave radiation emitted at ground level; D, longwave radiation leaving the Earth after attenuation by the atmosphere. Dark areas in the box show the location of absorption lines of O_3, H_2O and CO_2. Note that both axes are logarithmic. (Adapted from Gates [1].)

does when the atmosphere contains ice crystals or water droplets. Other particles in the atmosphere include dust blown from deserts and emitted by volcanoes. The smaller particles stay in the atmosphere for many months and cause a reduction in the direct radiation, but an increase in the diffuse component. Overall, these tend to cancel each other out.

A few meteorological stations measure diffuse and direct components of the total shortwave radiation. To do this, two sensors are employed, one of them being screened from the direct solar beam by a shade ring that must be adjusted from time to time, to take account of the seasonal progression of the sun's path in the sky. Comparison of the signal from the two sensors provides a measure of the direct component of radiation. Information on the separate contributions of direct and diffuse radiation are valuable in many kinds of problem involving radiation interception by surfaces of complex geometry. These problems occur in diverse fields of inquiry, from that of the plant scientist who may be interested in leaf canopies, to the architect who needs to know how much light will pass through a window and illuminate a room. Rough estimates of direct and diffuse radiation under various conditions can be obtained from Fig. 1.2.

A large proportion of the sun's rays are reflected back to space without heating the Earth's surface. The reflected proportion, known as the albedo, depends on the landscape and varies from 0.85 for fresh snow to less than 0.1 for water.

The rate at which total solar radiation is received on a horizontal surface near the ground may be about 1.0 kW m^{-2} at noon in the lower latitudes, and not much less than this in temperate latitudes when the sun is shining. However, when averaged over the whole year, including nights and winter, the mean value at Bracknell, England is only about 115 W m^{-2}.

The daily total of radiation is greatly affected by daylength as well as solar angle, so that northern latitudes receive substantially more solar energy than the tropics in summer months.

The relationship between latitude, time of year, solar angle and daylength can conveniently be shown as a solar track, in which the sun's path is shown on a map of the sky (Fig. 1.3). (For accurate use, corrections are required to take into account the variations in solar declination and local solar time [2].) Such solar tracks are useful when attempting to evaluate the influence of topography on the receipt of sunshine by plants, and have been used, in conjunction with photographs taken inside canopies of leaves, to study light interception by vegetation.

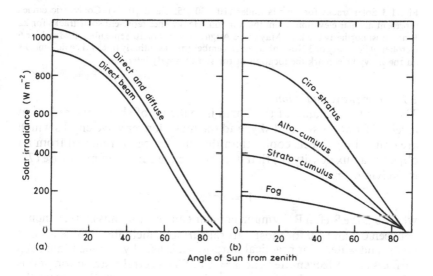

Fig. 1.2 Solar irradiance for (a) clear, and (b) clouded skies, as a function of solar angle measured from the zenith. As solar irradiance also depends on the aerosol content of the atmosphere, this relationship is approximate only. (Drawn from data in Gates [1].)

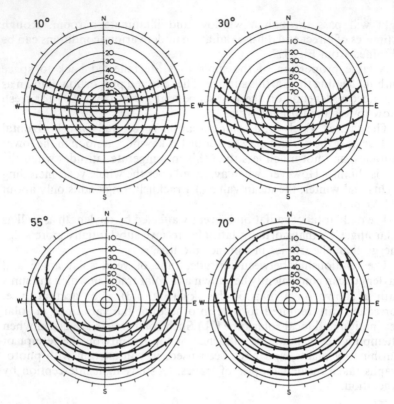

Fig. 1.3 Solar tracks for four latitudes (10°, 30°, 55°, and 70° N). Concentric circles represent angular elevations from the horizon. Heavy lines represent solar tracks for 22 June (most northerly track), 1 May or 12 August, 3 April or 10 September, 8 March or 6 October, 9 February or 3 November, 22 December (most southerly track). The tick-marks on the heavy lines mark the location of the sun at hourly intervals.

1.1.2 Longwave radiation

The radiation balance of the Earth is made up of shortwave radiation received from the sun and sky, and the much longer wavelengths which are emitted by objects cooler than the sun. Energy is radiated from an object at a flux density (**F**) that depends on the surface temperature (T_s) in Kelvin degrees:

$$\mathbf{F} = \varepsilon\sigma\,(T_s)^4$$

where σ is the Stefan Boltzmann constant and ε is the emissivity, which is between 0.9 and 1.0 for most natural surfaces, including leaves, animal coats, and water. For practical purposes it is usually assumed that such surfaces are perfect emitters, i.e. $\varepsilon = 1.0$. The spectral distribution of this energy depends on the temperature, but at around 280 K most of the energy is in the waveband 5–25 μm (Fig. 1.1). There are several important absorption bands within this range, especially those of H_2O

10

and CO_2. Hence, the atmosphere absorbs a proportion of the outgoing radiation, depending on such factors as atmospheric humidity and cloud cover. Cloud cover acts as a variable blanket in preventing the escape to space of much of the outgoing radiation.

Glass is opaque at these wavelengths, and so glasshouses are radiation traps, letting in shortwave radiation, but not letting out the longwave radiation emitted by the plants and other surfaces inside.

The sky itself is a source of longwave radiation, and may be said to have an apparent surface temperature. In reality this varies from place to place in the sky, but as a rule of thumb the clear sky has a surface temperature about 20 K lower than the air temperature. When overcast the difference between sky and air temperature is variable and much less than this.

1.1.3 Radiation balance

Both short and longwave radiations are thus received from above, while shortwave (by reflection) and longwave (by emission) are radiated from the vegetation back into space. The difference between incoming and outgoing radiation, taking all wavelengths together, is called net radiation. In the daytime, net radiation is positive; at night it becomes negative. Net radiation is generally measured using a net radiometer, an instrument consisting of two blackened thermopiles, one facing up and the other facing down (Fig. 1.4). The thermopiles are protected by a thin polythene film which transmits all wavelengths and prevents loss of heat by convection. They are connected electrically so that the temperature difference between the upper and lower surface is measured, and they are usually calibrated by placing the sensor between two large blackened plates whose surface temperatures and hence radiation emissions are known accurately and can be varied.

In a completely dry system, such as a concrete landscape or desert, the net radiation warms the air and ground causing an increase in temperature associated with the storage of heat, known as sensible heat. In a wet system a proportion of the incoming radiation is consumed in evaporation, and this is known as latent heat. For vegetated surfaces the partitioning of net radiation between latent and sensible heat components depends on whether the leaves and soil are wet and the extent to which the stomata are open. The ratio of sensible to latent heat, called the Bowen ratio, has been determined for various vegetation types and is found to depend on the characteristics of the vegetation and its physiological state (Fig. 1.5).

1.1.4 Units and terminology

There is still considerable confusion in the literature with regard to units. No attempt will be made to define all possible radiometric and photometric units as this is done at length elsewhere [4, 5].

The rate at which radiant energy is received by a surface is termed *irradiance*. The preferred unit is $J m^{-2} s^{-1}$ or (since $W = J s^{-1}$), simply

11

Fig. 1.4 Solarimeters and radiometers. (a) Moll–Gorczynski pyranometer, (Kipp and Zonen, Holland), usually called the Kipp solarimeter. (b) the Kipp solarimeter equipped with a shade ring to measure the diffuse component of solar radiation – note the shadow cast over the sensor (a correction should be made to allow for the small proportion of the diffuse radiation which is also blocked by the ring). (c) net radiometer (Swisstecho, Melbourne, Australia), in which the two sensing surfaces are protected from the wind by an inflatable dome of polythene which is transparent at all wavelengths. (d) a miniature net radiometer, especially useful for energy balance studies over patches of vegetation or large leaves, as the sensor is small and casts only a small shadow. (Photos: the author and D. Haswell.)

W m^{-2}. Irradiance may be measured on a surface at any angle, though by convention it is usually measured on a horizontal surface. In the absence of a surface, as in fluxes over vegetation, one is dealing with the *flux density* of radiant energy. The unit is still W m^{-2}.

In studies of physiological processes it is often more appropriate to use photometric units, in which the fundamental unit is not the Joule, but the photon or quantum. According to Planck's quantum theory, radiation is transferred in discrete packets called quanta (or photons if in the visible part of the spectrum). The energy transferred (**E**) is

12

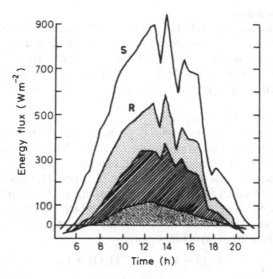

Fig. 1.5 Typical diurnal patterns of energy transfer over grassland: **S**, incoming shortwave radiation; **R**, net downward flux of all-wave radiation, partitioned between evaporation λE (light stipple), convection **C** (shading) and conduction to the ground **G** (heavy stipple) [3].

proportional to the frequency (f) of the radiation: $\mathbf{E} = hf$, where h is Planck's constant (Table 1.1).

The unit photons $m^{-2}\ s^{-1}$ is not preferred, as inconveniently large numbers occur in ordinary daylight. Instead, by defining a mole as Avogadro's number of photons (6.02×10^{23}), we may use the unit mol $m^{-2}\ s^{-1}$. Some light sensors are calibrated in Einstein $m^{-2}\ s^{-1}$, where one Einstein is 6.02×10^{23} quanta, but as there are two alternative and completely different definitions of the Einstein in popular use, this unit is deprecated by Incoll *et al.* [5].

Photon flux density is often measured with a sensor whose spectral sensitivity is matched to the broad waveband appropriate to photosynthesis, i.e. 400–750 nm. Then the quantity is termed photosynthetically active radiation or *PAR*. Ordinary daylight falls in the range 0–2000 mol $m^{-2}\ s^{-1}$.

Photometric and radiometric units cannot be readily interconverted, as (since $\mathbf{E} = hf$) the conversion factor depends on the spectral composition of the light source.

Table 1.1 Energy per quantum in the visible part of the spectrum

Wavelength (nm)	400	500	600	700
Frequency (s^{-1})	7.5×10^{14}	6×10^{14}	5×10^{14}	4.5×10^{14}
Energy per quantum (J)	5×10^{-19}	4×10^{-19}	3.3×10^{-19}	2.9×10^{-19}

1.2 The atmosphere

The composition of the Earth's atmosphere near the ground is well known (Table 1.2). Compared with that of neighbouring planets its unusual feature is the presence of much oxygen. In relation to the aqueous medium, in which life began, there are profound differences in density, viscosity and gas diffusion, as well as in composition.

1.2.1 History of the atmosphere

Soon after the Earth's formation, perhaps 4.5×10^9 years ago, the planet probably had no atmosphere or water. Volcanic emissions are thought to have provided an early atmosphere containing much water vapour, carbon dioxide and a few percent nitrogen, H_2S and SO_2. Water vapour condensed to form the oceans, leaving a CO_2-rich atmosphere with no oxygen. The oxygen component, which makes our planet so different from Mars and Venus, is now widely believed to have originated from photosynthesis:

$$H_2O + CO_2 \rightarrow (CH_2O) + O_2$$

When plants die and decompose the equation is reversed, and the oxygen originally produced becomes depleted. However, a small fraction of the carbon atoms fixed by photosynthesis escape oxidation by becoming buried as coal, oil and natural gas. Hence, a net oxygen enrichment of the atmosphere occurs at a rate proportional to the rate of formation of 'fossil fuel'. It has been pointed out that the burning of fossil fuel undoes the work of photosynthesis, and that if all fossil fuel were to be exploited then our oxygen would be depleted, and we would return to a CO_2-rich atmosphere.

Carbon dioxide has also been fixed in the sea, where, as carbonic acid it has been combined with calcium to form the shells of sea animals, eventually being turned into carbonate rocks. Carbon in this form considerably exceeds the carbon in fossil fuels (Table 1.3).

Table 1.2 The main gaseous constituents of the Earth's atmosphere below 100 km, percentage by volume, and a comparison with Venus and Mars

	Earth	Venus	Mars
Nitrogen (N_2)	78.08	1.9	2.7
Oxygen (O_2)	20.95		0.1
Carbon dioxide (CO_2)	0.03	98	95
Argon (A)	0.93	0.1	2.0
Water vapour (H_2O)	0–4		
Neon (Ne)	trace		
Helium (He)	trace		
Krypton (Kr)	trace		
Hydrogen (H_2)	trace		
Ozone (O_3)	trace		
Pressure (MPa)	0.1	9.0	0.0006

Table 1.3 Inventory of carbon near the Earth's surface (teratonnes [tera = 10^{12}])

Biosphere (marine)	0.01
Biosphere (non-marine)	0.01
Atmosphere	0.7
Ocean (dissolved CO_2)	40
Fossil fuel	8
Shales	8 000
Carbonate rocks	20 000

The burning of fossil fuels has undoubtedly caused an increase in the atmospheric concentration of carbon dioxide. The best measurements, made at Mauna Loa in Hawaii, show an increase over the period 1958–1975 from 315 cm^3 m^{-3} to 330 cm^3 m^{-3}, whilst earlier observations elsewhere suggest a concentration, a hundred years ago, as low as 290 cm^3 m^{-3} (Fig. 1.6). As carbon dioxide absorbs some of the outgoing longwave radiation, which would otherwise be lost to space, there has been much speculation as to whether the increase is causing an increase in global air temperatures. Estimates of this effect may be made from computer models which take into account such factors as the radiation properties of atmospheric gases and the dissolving of CO_2 in the oceans. Work in this field continues, but so far the estimated effect of a doubling in the CO_2 concentration has varied from only $+0.1°C$ to as much as $+4°C$ depending on the assumptions of the model concerned.

Another aspect of any increase in atmospheric carbon dioxide is that the rate of photosynthesis might be expected to increase. It is well known from laboratory studies that, within the limits concerned, the rate is proportional to carbon dioxide concentration as long as the leaves are brightly illuminated. Most leaves in a canopy are however, partially shaded, and these respond less to an increase in carbon dioxide. Overall,

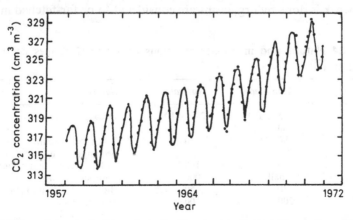

Fig. 1.6 Recent trends in the carbon dioxide concentration at Mauna Loa, Hawaii. (Keeling *et al.* [6].)

a doubling of the atmospheric carbon dioxide concentration might be expected to produce much less than a doubling in yield of agricultural crops, though, nevertheless, a substantial increase would occur. Of course, if global temperatures were to increase at the same time, the benefit for agriculture in northern climates could be immense.

There are several other gases that occur in trace amounts in the atmosphere. Some of them have significant absorptions in the solar or terrestrial spectrum, including HDO (heavy water), nitrous oxide, methane, carbon monoxide and ammonia. Many trace gases are familiar as pollutants, though it is not easy to distinguish the contribution that constitutes man-made pollution from that which is characteristic of clean air (Table 1.4). Several of the trace gases participate in natural cycles which, although of great importance, are not well understood [8].

Small particles, known as aerosols, are also constituents of 'clean' air. They include material from volcanoes, soil and salt from sea-spray, as well as pollens and spores.

1.2.2 Gaia hypothesis

Attention has been drawn to the disequilibrium among the gases in the Earth's atmosphere, especially when compared with that of Mars and Venus. The simultaneous occurrence of O_2 and CH_4 at their present concentrations, and the presence of a large proportion of the nitrogen as N_2 in the air instead of NO_3 in the oceans, are both cases of this disequilibrium [9]. It has been suggested that this state is maintained only as long as work is done by organisms, and that the biosphere as a whole might be considered as an active control system maintaining atmospheric homeostasis [9]. The notion has been called the 'Gaia' hypothesis, after the Greek word which (roughly translated) means 'Mother Earth' [10]. To many biologists these views, advanced by Lovelock, will seem teleological, and some of his examples of exactly how work is done by organisms are considered to be far-fetched in the

Table 1.4 Natural and man-made emissions on Earth (Mtonnes per year) (Almquist [7])

Pollutant	Man-made	Natural	Source
CO_2	$4-5 \times 10^3$	$2-3 \times 10^4$	Decay, oceans
CO	250	10^3	Fire
SO_2	80	50	Volcanoes, oceans, decomposition
H_2S	3	100	Volcanoes, decomposition
O_3	Small	2000	Photochemical
NO_x	30	550	Biological action in soil
Particles	200	1100	Sea salt, soil, volcanoes, fire
Heat	6.6×10^9 J yr^{-1}	5.5×10^{24} J yr^{-1}	Solar

extreme, but the general hypothesis – that the composition of the atmosphere is to a large extent caused by the activities of organisms, and that future stability depends on homeostasis achieved through living processes – is thought-provoking and deserves attention.

1.2.3 Atmospheric motion

The vertical transport of heat, carbon dioxide and water vapour between vegetation and the atmosphere depends on air movement. When air flows over a rough surface such as vegetation, surface layers are retarded by friction, giving rise to a region of locally-reduced wind speed, called the boundary layer. Unless the air is flowing very slowly over a very smooth surface, the stresses set up between adjacent air layers are sufficient to break up the flow, causing chaotic motion of the air which is said to be turbulent (Fig. 1.7). In turbulence, parcels of air (called eddies) moving at random, transport heat, carbon dioxide and water vapour from regions of high concentration to regions of low concentration, the overall process being called turbulent diffusion or simply turbulent transport.

The flux, F, of any entity in the vertical direction is proportional to the concentration gradient $\partial \chi / \partial z$:

$$F = -K \cdot \partial \chi / \partial z$$

the constant of proportionality is K the turbulent transfer coefficient. (Note the usual convention in micrometeorology to use z for vertical distance and χ for concentration.) In *laminar* boundary layers vertical transport depends on *molecular* diffusion and is much slower, the constant of proportionality then being D, the molecular diffusivity. The latter depends on the entity being transported and on the fluid involved: for CO_2 in air at 20°C, its value is about 1.5×10^{-5} m^2 s^{-1}. In contrast, K is more or less independent of the entity being transported and depends on the characteristics of the turbulence. Near leaves where the wind speeds are low and the eddies small, K might lie between 10^{-4} and 10^{-1} m^2 s^{-1}. Above the vegetation, where wind speeds are high and eddies large, K is always much higher.

In studies of leaves we are concerned with the boundary layer that develops immediately around the leaf surface, which has a thickness of

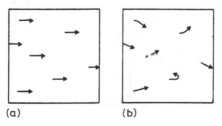

(a)   (b)

Fig. 1.7 (a) Laminar and (b) turbulent flow. Chaotic motion occurs in turbulent flow, causing vertical as well as horizontal transport of any substance entrained in the flow.

just millimetres. In micrometeorological studies of crops, measurements are made in the boundary layer that develops in the zone 1–10 m over the vegetation. Meteorologists on the other hand are concerned with the planetary boundary layer, up to a kilometre or so in depth, that is a characteristic of the entire landscape (Fig. 1.8). An important difference between these regions is in the size of eddies, or the scale of turbulence: the average size of eddy, and the extent of mixing and magnitude of K, increases with height above the ground.

1.2.4 Vertical distribution of variables

The lowest 10 km of the atmosphere, called the troposphere, contains 75% of all the gases and nearly all the water vapour. It is in the lowest part of this zone that most organisms occur.

Throughout the troposphere there are well-defined gradients of pressure, radiation and temperature. Even in the lowest hills the effect of the gradient in temperature is apparent, but it is only in the larger ranges of mountains that atmospheric pressure and solar radiation vary enough to have a significant effect on plant life (Fig. 1.9). In tall mountains like the Alps, direct solar radiation increases with altitude

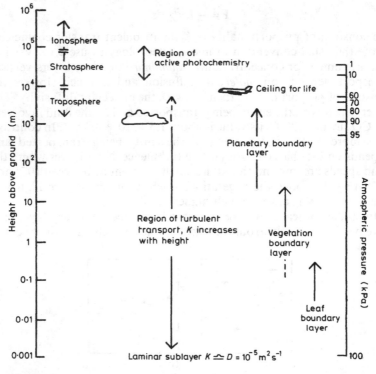

Fig. 1.8 Regions of the atmosphere showing extent of boundary layers and the region of turbulent transport. Note that the vertical scale is logarithmic.

18

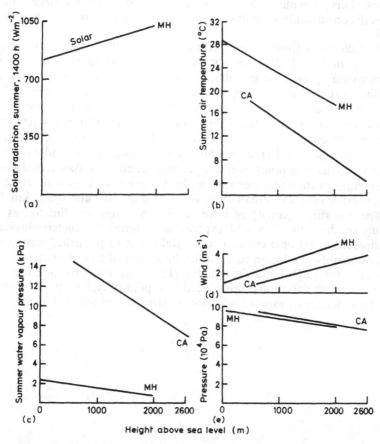

Fig. 1.9 Effect of altitude on principal environmental variables in the Central Alps (CA) and Mount Hermon, Eastern Mediterranean (MH). Absolute values are not strictly comparable because data were averaged over different periods, and in some cases values were calculated not measured. (From Körner and Mayr [11] and Cohen *et al.* [12].)

because of the smaller mass of air through which the sun's rays shine. Extremely high values of solar irradiance for brief periods have been reported in the Alps (Turner's data, quoted in Tranquillini [13]), even exceeding the solar constant. Such values occur when radiation is reflected from nearby ground and then scattered by clouds, so that a proportion that would otherwise be lost to space, is received at the recording station. The high albedo of snow accentuates this effect.

In Britain, the mountains are not very high, and the small increase in solar radiation that might be expected as a result of the reduced air mass, is more than compensated for by the formation of clouds on hill tops. Consequently, average solar radiation declines with altitude [14].

The decline in pressure with altitude implies a corresponding decline in the partial pressures of carbon dioxide, oxygen and other constituent gases. This determines the absolute altitudinal limit for organisms, though considerable adaptation is possible in plants as well as in animals [15].

The effect of altitude on temperature can be explained by considering a rising air parcel. As it rises it experiences a fall in pressure and a corresponding increase in volume, which must be accompanied by a decline in temperature. For every parcel of air that rises, another falls, experiencing an increase in pressure and so an increase in temperature. The overall effect is that temperature declines with altitude. For dry air, the rate of change is calculated as $10°C$ km^{-1} and is called the dry adiabatic lapse rate. In practice, air contains water vapour and if the air is cooled to its 'dew point', some of the water condenses. Associated with the change in state of water vapour, heat is released. Consequently the actual lapse rate is less than $10°C$ km^{-1}, and is often about $5°C$ km^{-1}.

There is still a paucity of information on mountain climates, as a result of the difficulties and expense of maintaining meteorological equipment in extreme environments. Reference to published measurements, mainly in relation to the altitudinal limit of tree growth, can be found in Grace [16] and Tranquillini [13]. In addition to the effect of altitude, much variation is introduced by topography, largely as a result of the influence of aspect and slope on the local energy balance.

2 Radiation coupling

> '. . . if the Sun's light consisted of but one sort of
> rays, there would be but one colour in the whole
> world, nor would it be possible to produce any new
> colour by reflections and refractions.'
> Newton

2.1 Introduction

The response of vegetation to changes in the atmosphere can be considered in terms of *coupling*. Two processes are said to be coupled if a change in one of them causes a change in the other. In a very general sense, living processes of all kinds are coupled to the atmosphere as all depend on an energy supply which is determined by atmospheric conditions. More specifically, we can identify two sorts of coupling:

1. Radiative coupling in which energy is transferred as electromagnetic vibrations.
2. Coupling across boundary layers, in which heat, carbon dioxide and water vapour are transferred by turbulent transport and molecular diffusion.

In this chapter radiative coupling will be discussed.

2.1.1 Spectral properties of leaves

It may be no coincidence that the human eye is most sensitive to green light ($\lambda \simeq 550$ nm). Leaves transmit and reflect most strongly in this part of the spectrum and the eye is apparently adapted to resolve the small differences that exist between the optical properties of leaves of different species. Man's ability to distinguish plant species from one another on their vegetative characteristics, and so recognize edible, medicinal and poisonous plants, must have been of considerable survival value in pre-agricultural food-gathering societies and so good green-vision may have been fixed by the process of natural selection [17].

Spectral properties of a typical leaf are shown in Fig. 2.1. In the visible part of the spectrum there is a high absorption with a minimum between 500 and 600 nm. Spectral properties in this part of the spectrum are determined by chlorophylls *a* and *b* (which have strong absorption peaks in the blue and red), by other pigments such as carotene and xanthophyll, and by the presence of structural material in the leaf which is nearly opaque. In the ultraviolet part of the spectrum, there is very strong absorption caused by numerous molecular species, including water. Inside the leaf, radiation is scattered by organelles, and so each beam entering the leaf has a rather long path length. The net effect of this is that the absorption is greatly increased and the absorption bands characteristic of each absorbing substance are broadened. Also much of

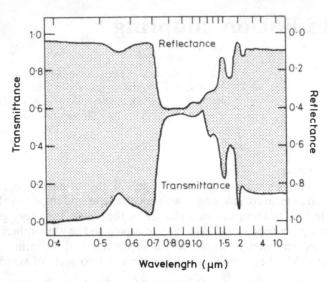

Fig. 2.1 Transmittance, reflectance and absorptance of a typical leaf. Stipple denotes absorptance.

the scattered radiation is back-scattered, appearing as diffuse reflectance. Another component of the reflected radiation is that which results from the interaction of radiation with the fine structure of the leaf surface, in particular, with the epicuticular waxes and any pubescence.

In the near infra-red there is an abrupt decline in the absorption of radiation by the leaf, in the range 700–1200 nm, accompanied by an increase in reflectance and transmittance. Nearly half of the incoming solar radiation is in this region. At higher wavelengths, liquid water is a strong absorber and the leaf is practically opaque, reflecting and transmitting almost nothing. Typical figures for the radiation absorbed by a leaf are shown in Fig. 2.2.

It is particularly interesting to know the extent to which the optical properties of leaves display ecological adaptations. In relation to this question, much work on the spectral properties of leaves has been published by Gates and his colleagues [1]. Gates has shown that many desert plants show especially high reflectance in the waveband 0.7–1.2 μm, substantially reducing the absorptance of solar radiation, a feature of presumed selective advantage in the desert. Many cacti are covered by a dense mat of thorns or hairs, which reflect visible and ultraviolet parts of the spectrum, and this may also be regarded as an adaptive feature. Gates [1] reviews experiments on pubescence, in which comparisons have been made between leaves in a naturally-pubescent state and those in which the hairs have been shaved off. The velvet plant *Gynura aurantiaca*, when shaved of its hairs, displays practically no change in reflectance in the visible part of the spectrum, but a significant decrease in reflectance does occur at 700–1000 nm. Above this wave-

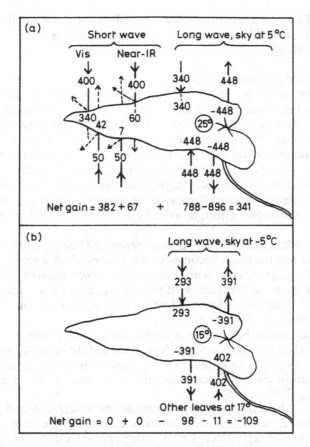

Fig. 2.2 Typical energy fluxes at a leaf in (a) bright sun, and (b) at night, assuming a clear sky and with other conditions as shown on the diagram. Units: W m⁻².

length however, the effect is reversed. Gates points out the difficulties in the interpretation of such experiments, as there are so many other possible functions of plant hairs [1]. One such function may be as a form of defence against herbivorous invertebrates.

Other studies have drawn attention to the change in leaf properties that occur along a natural gradient of increasing solar radiation. Thomas and Barber [18] collected *Eucalyptus urnigera* leaves from between 560 and 1050 m at Mount Wellington in Tasmania. The leaves differed in the extent of their surface waxiness, and displayed much more reflectance when collected at the higher altitudes, both in the visible and near infra-red parts of the spectrum. In the waveband 350–1350 nm the increase in reflectance was from 0.29 to 0.41, enough to substantially influence the energy balance of the leaf. However, the selective advantage of this feature is not clear, as the range of altitude is not really enough to cause much variation in solar insolation. The authors do, however, mention that the habitat at the high altitudes is more open and seedlings grow on boulders and among rocks with very high reflectivity.

2.1.2 Optical properties of canopies

Differences in the properties of fully grown agricultural crops, with respect to their optical properties, are rather small [17]. Large differences do occur between crops in different stages of growth, when differing amounts of bare ground are exposed. For other vegetation types, where a greater range of plant form and leaf type is evident, the differences are greater. Tall vegetation, especially coniferous forest, generally has a low solar reflectivity as a result of its high cavity depth and light scattering by needles, features which together cause trapping of radiation – sometimes called the 'velvet-pile' effect, as velvet appears rich black for the same reason.

Reflectance of almost every surface increases when the angle between the surface and the beam becomes small. Leaves which approximate in form to a cylinder interact with sunlight in a rather different manner to broad flat leaves, as irrespective of their angle, a significant proportion of incident sunlight strikes the leaf surface at a small angle and is scattered in many directions.

Most vegetation behaves as a 'compound surface' reflecting more radiation when the angle between the surface and the beam is small, as in the early morning and late evening, and this effect is especially pronounced in those surfaces that are composed of near-horizontal leaves (Fig. 2.3).

There have been many attempts to model the penetration of light into plant canopies, as a first step towards calculating the rate of crop photosynthesis. Such models are necessarily complex if they are to be at all realistic, as factors including the distribution of leaf angles with height, the dispersion of leaves in space and the geometry of solar radiation must all be taken into account. The main advantage of a

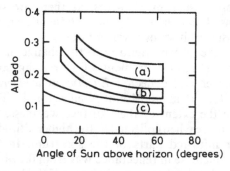

Fig. 2.3 Albedo (shortwave reflectance) as a function of solar angle. Envelopes show the range of measurements made in different kinds of vegetation when fully grown: (a) herbaceous; (b) deciduous forest; (c) coniferous forest. (From several sources.)

24

modelling approach is that, once the model is complete and adequately tested by comparison with measurements in the field, it does enable a systematic assessment of each factor in turn. This is obviously an advantage to crop breeders and agronomists who require as an operational target the specification for the optimum canopy [19]. However, the approach has so far been much less useful to ecologists in their attempts to understand the complex relationships between species in a mixed stand of vegetation. Even foresters, who after all deal with simple monocultures for much of the time, have yet to find practical uses for models of light penetration.

Detailed descriptions of models of light penetration can be found in Monteith [20], de Wit [21], Cowan [22], Norman and Jarvis [23] and Norman [24]. Such models can be used to estimate penetration of infra-red as well as visible components of the radiation as long as data are available on the spectral properties of the leaves throughout the relevant waveband (Fig. 2.4). Attempts to make models more realistic necessitate considerable mathematical complexity, especially if it is intended to simulate sunflecks (patches of the forest floor illuminated by direct sunlight), and to take into account the effect of non-random distribution of foliage.

Thus, all mathematical treatments of light interception must involve a degree of simplification. It has been known for a long time that leaves move in response to changes in the light regime. Actually, rather few of the complex phototropic or heliotropic responses that many species possess have been adequately described. In *Helianthus annuus*, the sunflower, these responses are pronounced in the developing inflorescence as well as in the leaves. Both show diurnal movements in the

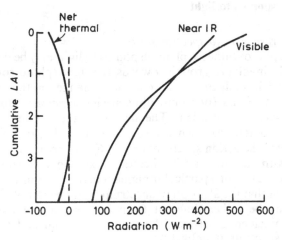

Fig. 2.4 Mean profiles of downward visible, downward near infra-red and net thermal radiation in a canopy with leaves facing all directions and with a leaf reflectance and transmittance of 0.1 in the visible and 0.4 in the near infra-red. (Calculation by Norman [24].)

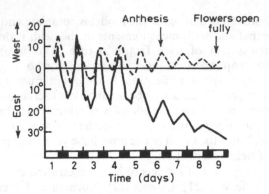

Fig. 2.5 East–west angular motion of sunflower leaves (broken lines) and inflorescences (solid lines). Black zones on abscissa denote night. (From Lang and Begg [25].)

horizontal plane between east and west, tracking the sun. The leaves move in the vertical plane to match the solar angle, which has the effect of increasing the interception of solar energy. When the flowers open, their orientation becomes fixed to the east (Fig. 2.5). It is said that this is an adaptive feature: it reduces the floral temperature at mid-day, which might otherwise be lethal; and the dew dries more quickly in the early morning sun, hence reducing the chance of fungal attack [25]. In some other species, leaf movements have the effect of *reducing* solar radiation intercepted, perhaps avoiding photodestruction of chlorophyll in very bright light or reducing leaf temperature [1].

2.2 Plant responses to light

2.2.1 Plant responses to spectral shifts
Apart from photosynthesis, plant responses to light may be divided into those simply involving growth towards the light (phototropism) and those in which a distinct and complex change in the pattern of differentiation occurs (photomorphogenesis). Responses to light are usually wavelength dependent. The amount of the response illicited by different wavelengths can be measured experimentally, and the resulting graph is called the action spectrum (Fig. 2.6). Comparison of an action spectrum with the absorption spectra of known pigments provides evidence for the role of a particular pigment in that response (although it is well known that the absorption spectrum of an extract of a particular pigment may be rather different from its spectrum *in vivo*, and so complete coincidence between the action spectrum and the *in vitro* absorption spectrum is unlikely).

The action spectrum of most phototropic responses suggests the involvement of carotenoids or riboflavin, whilst that of photomorphogenesis is suggestive of the pigment phytochrome (Fig. 2.6).

Phytochrome is a large protein molecule existing in two isomeric forms which are interconvertable:

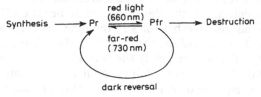

Fig. 2.6 Action spectra of plant responses to light and the main absorption bands of the corresponding pigments. The spectrum of human vision is included for comparison.

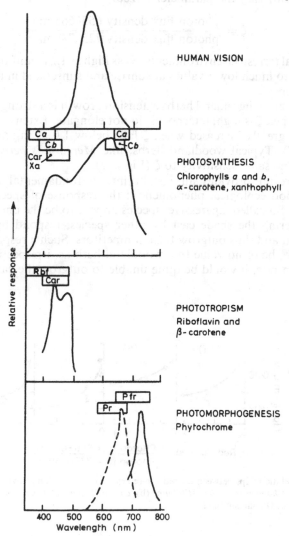

The ratio of Pr to Pfr is determined by the relative amounts of red and far-red in the incident radiation. As all leaves absorb strongly at 660 nm but only weakly at 730 nm, natural shade is enriched in far-red radiation and so may be detected by the plant as a shift in the ratio of Pr to Pfr [26]. Phytochrome has many other roles in the plant, apart from the direct sensing of shade: these include the detection of daylength (through the dark reversal to Pr), and probably the detection of the depth of soil by some germinating seeds. For a full account of phytochrome, see [27].

The growth response of plants to the quality of radiation has been investigated in special growth cabinets in which the photosynthetically-active radiation is kept constant whilst altering the amounts of far-red radiation to vary the parameter ζ (zeta):

$$\zeta = \frac{\text{photon flux density 655–665 nm}}{\text{photon flux density 725–735 nm}}$$

In natural terrestrial environments ζ is as high as 1.2 in mid-day sunlight, but falls to much lower values at sunrise and sunset and in the shade of vegetation.

In fast growing ruderal herbs extension growth is a strong function of ζ [26]. When ζ is high, internodes do not elongate. Extension growth is however greatly increased when ζ falls below 1.0 as in natural shade (Fig. 2.7). Typical woodland herbs, like *Mercurialis perennis* do not display the same sensitivity to ζ (Fig. 2.7).

This result provides an insight into a fundamental and poorly understood ecological phenomenon: the response of species to competition. So-called 'aggressive' species appear to be the ones which, on encountering the shade cast by other species, respond by internode extension and thus outgrow their competitors. Such a response would, of course, be of no value to a woodland herb as, irrespective of a fast extension rate, it would be quite unable to outgrow the canopy of tall trees.

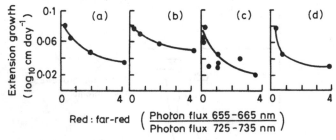

Fig. 2.7 Relationships between the red:far-red ratio and extension growth in four ruderal herbs: (a) *Chamaenerion angustifolium*; (b) *Urtica dioica*; (c) *Chenopodium album*; (d) *Sinapis alba*. (From Smith [26].)

The photomorphogenic response of plants to shade includes a wide range of developmental changes apart from those involved in stem extension. There are profound changes in the photosynthetic machinery, especially in the light-harvesting system, and changes in the leaf area to weight ratio. It seems likely that many of these changes are triggered not by the overall reduction in photon flux, but by the spectral changes that are associated with natural shade. On the other hand, it is possible to demonstrate that plant development does respond to the quantity of photons as well as the quality. Smith [26] showed that *Chenopodium album* responded to low-ζ shade by producing the maximum extension rate at the expense of leaf development, whereas neutral shade produced maximum leaf development at the expense of stem material. It is not known whether the plant detects neutral shade by somehow measuring the rate at which photosynthesis proceeds, or by means of a special pigment system. Smith [26] points out that some fungi detect minute differences in the rate of arrival of photons on two sides of a sporangiophore, and suggests that plants may have the capacity somehow to count the arrival of photons.

2.2.2 Photosynthetic machinery

Photosynthesis depends on (a) the capture of photons by chlorophyll to provide ATP and reducing power, and (b) the subsequent utilization of these products to drive the chemical reactions that make carbohydrates from carbon dioxide. These two parts of photosynthesis, called the light and dark reaction respectively, are usually closely coupled to each other in the cell, although they can be separated for study *in vitro*. In the cell, photosynthesis occurs in saucer-shaped organelles called chloroplasts. They are 5 to 10 μm in diameter, and look like currant buns under the light microscope, due to the aggregation of chlorophyll-containing membranes into stacks called grana (Fig. 2.8).

The light reaction involves the very fast (10^{-15} s) photochemical response of chlorophyll and other pigments, which on absorbing a photon assume an excited state, in which electrons associated with the double bonds in the 'head' of the molecule are raised momentarily to one of two possible higher levels of energy. In a solution, excited chlorophyll molecules rapidly (10^{-9} s) decay to the ground state with the emission of a photon of slightly longer wavelength than the one absorbed. This emission appears as red fluorescence in chlorophyll solutions. In chloroplasts, molecules of chlorophyll are fixed to membranes and packed tight with the 'head' parts adjacent, so that energy contained in one excited molecule can be passed to another, which in turn can pass the energy on to yet another. The overall functional unit consists of hundreds of such molecules of chlorophyll *a* and *b*, acting as an antenna which captures the energy and relays it to a reaction centre. Since about 1960 it has been believed that there are two reaction centres within this functional unit, forming the heart of two collaborating photosystems

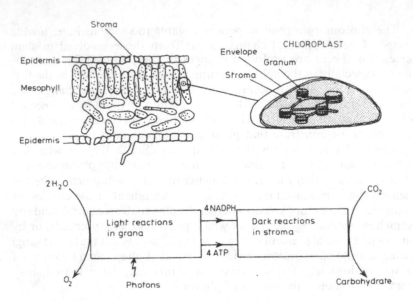

Fig. 2.8 Gross features of the photosynthetic system. The upper diagram shows the distribution of chloroplasts (small dots) in the mesophyll, with a much enlarged cut-away view of the chloroplast contents. The lower diagram summarizes the essential features of the photosynthetic process.

(Fig. 2.9). The 'Z scheme' originates from the hypothesis then put forward by Hill and Bendall, and has since been the subject of much research. Many of the details are not well understood even now.

The centre for photosystem 1 is connected to an antenna in which chlorophyll a predominates, and is most active at 700 nm. Chlorophyll b predominates in photosystem 2, which is most active at somewhat shorter wavelengths. Reaction centres are called P_{700} and P_{682} respectively, and consist of aggregates of chlorophyll and protein.

In photosystem 2, the energized electrons are passed to a series of electron carriers that utilize the energy to form ATP, constituting a sequence called *non-cyclic photophosphorylation*. (The expelled electrons are replaced by those released, together with protons and oxygen, as a result of the splitting of water. When photosystem 2 is isolated *in vitro* it evolves O_2 on illumination. It is this reaction that is the source of O_2 evolution in photosynthesis.) Photosystem 1 forms part of the electron pathway, giving the electron a final boost in energy to achieve the reduction of the iron–sulphur protein, ferredoxin. This substance provides reducing power to reduce NADP to NADPH, in the presence of a flavin enzyme. An alternative pathway of electron flow, *cyclic phosphorylation*, involves only photosystem 1. On illumination, the reaction centre expels an electron to reduce ferredoxin. Instead of the reduced ferredoxin passing its reducing energy to NADP, the electron is passed down a cytochrome chain, the energy being used to generate

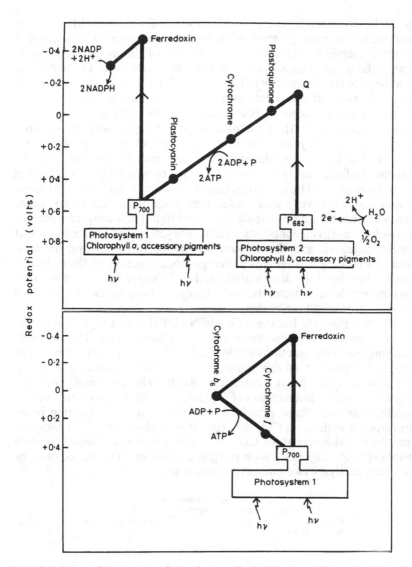

Fig. 2.9 Schemes of electron transfer in photosynthesis. The upper diagram is non-cyclic photophosphorylation and involves electron transfer from water to NADP, creating reducing power (NADPH) and chemical energy (ATP). The lower diagram is cyclic photophosphorylation in which the electrons come from chlorophyll, to which they eventually return. Cyclic photophosphorylation is known to occur when isolated chloroplasts are supplied with ferredoxin, but its contribution *in vivo* may be small. For further discussion see Leech [28] or Boardman [29].

ATP. Finally the electron is returned to the reaction centre.

The exact configuration of the photosystems depends partly on growing conditions, especially the light climate. Plants from the shade

31

generally possess a higher ratio of chlorophyll a : chlorophyll b, more membranes in each granum, with an irregular granum arrangement which presumably enables better collection of diffuse light. In shade plants there are generally more chlorophyll molecules to each P_{700}, enabling the reaction centre to be supplied from a larger antenna and so fed with energy at a fast rate even though the light is dim.

Other pigments occur in association with chlorophyll. Algae contain *accessory pigments* in their antennae, such as the phycoerythrins (in red algae), the phycocyanins (in blue–green algae) and the carotenoid fucoxanthol (in brown algae). These pigments enable the algae to harvest the light quanta over a wider range of wavelengths; red light is strongly attenuated by several metres of sea water and chlorophyll alone would not work very well under such conditions. There are other pigments that do not participate directly in light harvesting: carotenoids are widespread in higher plants, causing the yellow and brown colours of senescent leaves, and the colours of carrots, tomatoes and oranges. Mutants lacking carotenoids suffer photodestruction of their chlorophyll when brightly illuminated, and it is suggested therefore that carotenoids have a protective role, carrying away excess energy from excited chlorophyll molecules.

All green plants utilize the ATP and NADPH to fix CO_2 in a sequence of biochemical reactions known as the Calvin cycle. The enzymes carrying out these reactions are located in the chloroplast stroma. The first stable product of this sequence is a 3-carbon compound, phosphoglyceric acid. Most temperate plants rely exclusively on this sequence, and are referred to as C3 plants. In evolutionary terms the C3 condition is considered primitive: it works best when running in an atmosphere with more CO_2 and less oxygen than today's. It happens that the initial acceptor for CO_2, a 5-carbon compound called ribulose bisphosphate, may react with oxygen as well as CO_2, as the enzyme involved catalyses two competing reactions:

$$
\begin{array}{ccccc}
\text{(C5)} & + & \text{(C1)} & \xrightarrow{\substack{\text{RBP carboxylase}\\ \text{oxygenase}}} & \text{2 (C3)} \\
\text{Ribulose} & & CO_2 & & \text{Phosphoglycerate} \\
\text{bisphosphate} & & & & \downarrow \\
& & & & \text{(C6) Sugar}
\end{array}
\tag{2.1}
$$

$$
\begin{array}{cccccc}
\text{(C5)} & + & O_2 & \xrightarrow{\substack{\text{RBP carboxylase}\\ \text{oxygenase}}} & \text{(C3)} & + & \text{(C2)} \\
\text{Ribulose} & & & & \text{Phosphoglycerate} & & \text{Phosphoglycollate} \\
\text{bisphosphate} & & & & \downarrow & & \downarrow \\
& & & & \text{(C6) Sugar} & & \text{(C1) } CO_2
\end{array}
\tag{2.2}
$$

The phosphoglycollate formed in reaction (2.2) is involved in another sequence of reactions in which one third of its carbon is released as CO_2. This release of CO_2 during photosynthesis has long been recognized, and was originally termed photorespiration.

Reaction (2.2) exceeds reaction (2.1) at high temperatures, and so the photosynthetic rate of C3 plants usually declines at higher temperatures. Reaction (2.2) also predominates when CO_2 is in short supply at the carboxylation sites, as when leaves are brightly illuminated, hence exerting a 'brake' on photosynthesis at high light intensities.

Tropical grasses and certain dicotyledonous plants have, in addition, the C4 pathway, in which the first product is a 4-carbon compound, oxaloacetate. The enzyme catalysing the reaction is not subject to competitive effects involving O_2 and so the reaction does not slow down when CO_2 molecules become scarce:

$$\underset{\substack{\text{Phosphoenol-}\\\text{pyruvate}}}{\text{(C3)}} + \underset{CO_2}{\text{(C1)}} \xrightarrow{\text{PEP carboxylase}} \underset{\text{Oxaloacetate}}{\text{(C4)}} \longrightarrow \underset{\text{Malate}}{\text{(C4)}} \qquad (2.3)$$

The malate is broken into pyruvate and CO_2:

$$\underset{\text{Malate}}{\text{(C4)}} \longrightarrow \underset{\text{Pyruvate}}{\text{(C3)}} + \underset{CO_2}{\text{(C1)}} \qquad (2.4)$$

This decarboxylation occurs in a special tissue, the bundle-sheath, which occurs around the vascular bundles. The CO_2 liberated is then fixed by the Calvin cycle, which is restricted in most C4 species to the bundle-sheath chloroplasts.

Hence, in C4 plants, the Calvin cycle operates in a CO_2 enriched environment (as it would have done in a primitive atmosphere) and follows the fruitful reaction (2.1) instead of the apparently wasteful reaction (2.2). The rate of photosynthesis is thus potentially higher in C4 plants than in C3. As the light intensity is increased, and internal CO_2 is depleted, there is no switch to reaction (2.2) and so light-saturation does not occur so readily (Fig. 2.10).

PEP carboxylase also has an important CO_2-trapping role in members of the Crassulaceae, and many other succulents, as part of their adaptation to life in the desert. The stomata remain shut in the day and water is thus conserved. At night the stomata open and CO_2 is fixed to form malate which is stored in vacuoles. This does not consume much energy and so the photosystems are not required. In the day the stomata shut, but the photosystems continue to operate to supply the energy for the formation of C_6 compounds from malate. This sequence is called Crassulacean Acid Metabolism (CAM). Overall rates of photosynthesis are not high, but water is conserved. The similarity between CAM and C4 photosynthesis should be noted: in the former the first carboxylation is separated from the second by *location* within the tissues, in the latter the carboxylations are separated by *time of day*.

All of the partial reactions of photosynthesis are temperature-sensitive, with the exception of the initial capture of photons by chlorophyll, which is a purely photochemical event. There are important

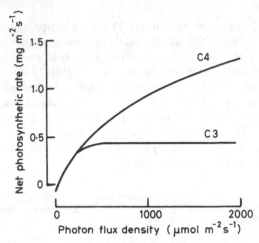

Fig. 2.10 Typical light response curves for C3 and C4 photosynthesis. Note that C4 plants have higher absolute rates of photosynthesis and that light saturation does not occur at ordinary photon flux densities.

differences between the overall temperature-sensitivity of C3 and C4 plants. As already mentioned, the affinity of ribulose bisphosphate carboxylase for CO_2 declines rapidly with increasing temperature, high temperatures favouring reaction (2.2) as opposed to reaction (2.1). Hence CO_2 becomes progressively rate limiting as the leaf temperature is increased, and it is this factor which largely determines the shape of the temperature response curve. Björkman [30] points out that stomatal conductance, as it affects the supply of CO_2 to the mesophyll, should also exert an influence on the optimum temperature of photosynthesis in C3 plants.

In general, C4 plants thus display much higher temperature optima for photosynthesis than do C3 plants, and are consequently a major constituent of tropical and subtropical vegetation. A few C4 plants do, however, manage to survive, and even thrive, in cooler climates – notably the grass *Spartina townsendii* and species in the dicotyledonous genera *Atriplex* and *Salsola* [31, 32].

2.2.3 Photosynthetic response to light
The rate of photosynthesis in C4 species continues to increase with increasing quantum flux, so that the highest rates are much higher than those achieved by C3 plants which photosaturate at relatively low levels of light (Fig. 2.10). Relationships like these are displayed by horizontal leaves in direct illumination. In a canopy of leaves, individuals may be inclined at any angle or orientation. It is important to realize that many leaves, even on a C3 plant, do not photosaturate as they are inclined to a small angle with the solar beam and so intercept only a small proportion of the sun's rays. Moreover, many of the leaves in a canopy are shaded and do not receive direct sunlight: such leaves

34

will not photosaturate even when the sunlight incident on the top of the canopy is maximal. Furthermore, on cloudy days light saturation may not be achieved even in leaves at the top of the canopy which are favourably displayed. Consequently, for all these reasons, photosynthesis on a 'per day' or 'per week' basis is often a linear function of the receipt of solar energy (Fig. 2.11), even in C3 plants.

2.2.4 Efficiency of photosynthesis

Under ideal conditions in the laboratory, eight quanta of energy are associated with the liberation of one O_2 molecule. The efficiency implied by an eight quantum process:

$$CO_2 + H_2O \xrightarrow{\text{8 quanta}} (CH_2O) + O_2$$

can be calculated by reference to Table 1.1, which shows that a quantum of energy in the photosynthetically active parts of the solar spectrum has, on average, an energy of 4.15×10^{-19} J. One gram molecule of glucose $(C_6H_{12}O_6)$ burned in a calorimeter yields 2809 kJ, so a sixth of it (CH_2O) is equivalent to 468 kJ. Dividing this by Avagodro's number we have the energy contained in one molecule of $(CH_2O) = 7.8 \times 10^{-19}$ J. This represents the amount of energy fixed every time eight quanta $(33.2 \times 10^{-19}$ J$)$ are absorbed. The efficiency is thus $7.8/33.2 \times 100\% = 23.5\%$. This is quite good by engineering standards, being rather better than most solar cells, but not as good as a steam engine. However, to achieve such favourable efficiency, carefully prepared algal solutions must be illuminated by dim light at just the right wavelengths. These conditions only exist in the laboratory; in field conditions much lower efficiency is observed. As we have seen, chlorophyll is capable of capturing only a fraction of the solar spectrum. Much sunlight strikes the ground, or non-photosynthetic parts of the vegetation. Much vegetation undergoes cycles of stress on a daily or annual basis, in which the photosynthetic system is under strain and does not function at its best.

Rates of biomass production in several systems are shown in Table 2.1. When measured over a whole year, the efficiency of photosynthesis, calculated in relation to the solar energy *incident* on the vegetation, is very low. The most productive crop, based on marine microalgae (a prototype energy plant used by Professor Wagener of the Technical

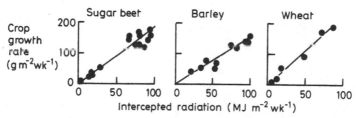

Fig. 2.11 Relationship between intercepted radiation and crop growth rate, for three temperate crops. (From Biscoe and Gallagher [33].)

University of Aachen, West Germany) is still poor in relation to the quantum efficiency of the photosynthetic machinery (Table 2.1).

Also included in Table 2.1 is an estimate of the land area required by one person for food, assuming a vegetable diet. In Great Britain, self sufficiency in food for a population of 60×10^6 people would require 2.4×10^{10} m^2 of a total land area of 2.3×10^{11} m^2, again assuming a population of vegetarians. On the other hand, any attempt to run all our power stations by biomass energy should be viewed with caution: the area needed to provide all our energy from biomass is more than the total land area. Other nations with plenty of land per capita can, however, look forward to a biomass contribution to their future energy requirements.

Table 2.1 Photosynthetic productivity and human life. Assumptions: 1 g dry weight yields 20 kJ on combustion; solar radiation in temperate conditions is 3.2×10^9 $J\ m^{-2}\ yr^{-1}$ and 4.0×10^9 $J\ m^{-2}\ yr^{-1}$ in the tropics; one person requires 8×10^9 J per year of food energy to sustain a normal life, and 1×10^{11} J per year to provide warmth and all benefits of industrialized society

| | Approximate productivity | | Efficiency (%) | Area required per person (m^2) for | |
	$(kg\ m^{-2}\ yr^{-1})$	$(MJ\ m^{-2}\ yr^{-1})$		food	energy
Temperate crops	1	20	0.6	400	5000
Temperate forest	1–2	20–40	0.6–1.2	—	2500–5000
Tropical forest plantation	2–6	40–120	1–3	—	830–2500
Marine microalgae at coastal desert	6–9	120–180	3–4.5	44–67	555–830

3 Coupling through boundary layers

That blueness is what pine-tips, weathered thus
And backed with pine-tips, make of air,
Region of compromise which they two share.
Norman MacCaig

3.1 Electrical analogues

In Chapter 1 it was mentioned that the transfer of materials between vegetation and the bulk of the atmosphere depends on the turbulent nature of the planetary boundary layer. If there was no such turbulence, a rapidly photosynthesizing stand of vegetation would quickly deplete the surrounding air of carbon dioxide. In fact, parcels of fresh air from above are continually mixing with the air in the vicinity of the vegetation. Sometimes turbulent transport is simply called turbulent diffusion, recognizing that the process resembles diffusion in that the net effect is to transfer material from places of high concentration to those of low concentration, but distinguishing it from molecular diffusion, which is orders of magnitude slower.

In the transfer of materials to and from plants, some exchanges do indeed occur by molecular diffusion, such as the passage of water vapour and carbon dioxide through stomata. Over the vegetation however, it is the motion of air parcels that is the vehicle for transport.

An analogy between diffusion in its widest sense and the flow of electricity may be drawn. Like diffusion, the flow of electrons from A to B along a conducting wire is proportional to the concentration difference between A and B, otherwise known as voltage or potential difference. Ohm's law for the flow of current:

$$\frac{\text{Current}}{\text{(amps)}} = \frac{\text{Potential difference (volts)}}{\text{Resistance (ohms)}}$$

is just like Fick's law for diffusion:

$$\frac{\text{Flux of diffusing gas}}{\text{(kg m}^{-2}\text{ s}^{-1}\text{)}} = \frac{\text{Concentration difference (kg m}^{-3}\text{)}}{\text{Diffusion resistance (s m}^{-1}\text{)}} \quad (3.1)$$

This similarity between diffusion and flow of current leads to the representation of the diffusion pathway as a chain or network of resistors (Fig. 3.1). Such a representation is often used by plant physiologists dealing with the flux of water vapour through stomata, and micrometeorologists considering fluxes by turbulent diffusion above the canopy. One advantage of this representation is that stomatal physiologists and micrometeorologists can have useful discussions in

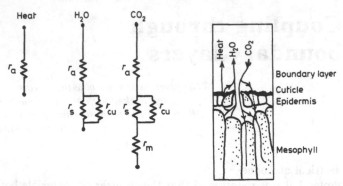

Fig. 3.1 Resistances for gas and heat exchange at the surface of a single leaf: boundary layer, r_a; stomatal, r_s; cuticular, r_{cu}; mesophyll or residual, r_m. The drawing on the right shows the pathways in relation to leaf structure.

the knowledge that they are working on parts of the same diffusion chain.

As in flows of electricity, rules of addition apply: two resistances, r_1 and r_2, joined end to end (in series) are added to find the overall resistance, R:

$$R = r_1 + r_2 \qquad (3.2)$$

whereas for resistances joined side by side (in parallel) the overall resistance, R is found from:

$$\frac{1}{R} = \frac{1}{r_1} + \frac{1}{r_2} \qquad (3.3)$$

For some purposes it is useful to use the reciprocal of resistance, conductance, G:

$$G = \frac{1}{r} \qquad (3.4)$$

Electrical analogues can be extended beyond networks of resistances to include capacitors to represent storage, as in the storage of heat by a tree trunk. They can also be used to represent water flow in the plant, which occurs at a rate proportional to differences in water potential divided by the hydraulic resistance of the conducting pathway. Some care with units is necessary, however, when connecting the resistance chain for liquid water within the plant to the resistance chain for water vapour exchange with the atmosphere.

Analysis of resistance pathways has been applied to very many diffusing entities, including pollutant gases and particulate matter suspended in the atmosphere. It has also been used in the related field of animal ecology to represent the diffusion of heat through animal fur.

3.2 Coupling through resistance chains

If the temperature of a leaf is nearly the same as that of the atmosphere,

38

and it responds immediately to a step-wise change in air temperature by a corresponding change in plant temperature, then the plant may be said to be *closely coupled* to the atmosphere [34]. Resistances for heat transfer would in this case be very small, and there would be no capacitors in the diffusion pathway. The temperature of a root tip on the other hand, would be very loosely coupled to the atmosphere, connected by a rather complex circuit of large resistors and a sizeable capacitor.

3.2.1 Boundary layers of leaves

When heat and mass exchange occurs between a leaf and the atmosphere by diffusion through the boundary layer, the process is called *forced convection* when the wind is blowing and *natural convection* in nearly still conditions. In the latter, the motion of air over the surface of the leaf occurs as a result of heating or cooling of the surface, leading to local changes in the density of the surrounding air. Warm air rises or cool air falls as result of these changes.

In forced convection, the flow of air around a smooth flat plate often forms a laminar boundary layer in which the lines of flow are parallel to each other. The mechanism for transport (normal to the surface) of gases and heat is that of molecular diffusion across the lines of flow (Fig. 3.2). The tendency for this smooth flow to break up into chaotic motion is given by the Reynolds number:

$$Re = \frac{ul}{v} \qquad (3.5)$$

where u is the wind speed, l is the length of surface over which the fluid has passed, and v is the kinematic viscosity of the fluid. In most ordinary conditions the Reynolds number over leaves is $0–10^4$ whereas, the critical Re required to precipitate turbulence in smooth airflows is generally accepted to be about 10^5.

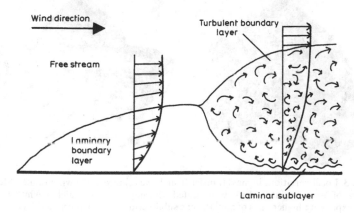

Fig. 3.2 Air flow over a smooth flat plate, showing transition from laminar to turbulent flow.

39

However, the air flow incident on leaves in nature is already turbulent, and the leaf is rarely smooth and flat. Structural details of leaves such as prominent veins and decurrent edges have the effect of tripping the air flow and causing turbulence at very low Reynolds numbers [35]. In wind tunnel studies it is possible to demonstrate that serrated edges on leaves also cause turbulence, each tooth generating a train of vortices which stream across the leaf from leading to trailing edge (Fig. 3.3).

Even in a turbulent boundary layer there is a zone (less than 1 mm thick) immediately next to the surface, in which the tendency of air to cling to the surface prevents turbulence. This zone is called the *viscous sublayer* or laminar sublayer (Fig. 3.4).

The average thickness of the boundary layer is related to leaf size. Thus small leaves have thin boundary layers which constitute small boundary layer resistances, and their temperatures are never very different from that of the surrounding air. Large leaves have thick boundary layers with large boundary layer resistances and temperatures which may differ substantially from that of the surrounding air. At high wind speeds the boundary layer is thinner than at low speeds and the resistance correspondingly smaller. At a critical wind speed a transition occurs from laminar to turbulent flow, causing an abrupt change in the relationship between wind speed and exchange rate, as an additional vertical component of transfer comes into play. The main determinants of boundary layer resistance are therefore leaf size and wind speed, with leaf form exerting a secondary effect through its effect on turbulence.

Boundary layer resistance may be evaluated by making a model of the leaf in question and measuring the rate of diffusion of any entity between the leaf and the air. The concentration difference between the leaf and the atmosphere must be known fairly accurately so that Equation (3.1)

Fig. 3.3 Local variations in mass transfer from leaves exposed in a wind tunnel. Metallic models of leaves, painted black, were coated with naphthalene, which is white. Certain areas experience higher rates of naphthalene sublimation, and so appear as black patches. Note the influence of serrated edges. The wind was from the left. (The author was aided by Alan Harper, of the Department of Chemical Engineering, Edinburgh University.)

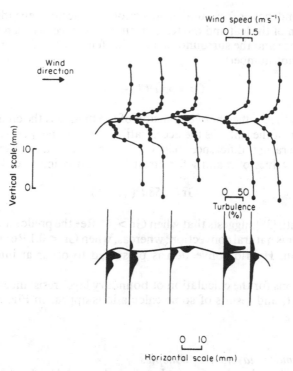

Fig. 3.4 Air flow over a real leaf. The thick black line is a transect of the leaf, showing roughness due to topography and veins. Only a part of the upper surface behaved like a flat plate. Elsewhere the boundary layer was turbulent [35]. Note that the laminar sublayer mentioned in Fig. 3.2 is not detected by this technique.

can be applied. The assumption is made that the aerodynamics of the model are the same as that of the real leaf; though inevitably, leaf minutae are missing from the model. There are several techniques in current use. An older method is to measure the evaporation rate of water from a paper model under specified conditions [36]. More accurate is the cooling curve technique in which the model is made of polished metal (near-zero emissivity) and the diffusing entity is heat [37].

There is much discussion in the literature about the estimation of boundary layer resistances, from formulae used in engineering sciences (engineers frequently require to estimate heat dissipation from surfaces, as for example, in the heat exchangers of power stations). Engineering formulae hold good for smooth flat surfaces in laminar flow; the results obtained for leaves in more realistic flows of air, or in the natural wind, suggest that the exchange rates for real leaves may be up to twice that calculated from the formulae, presumably because of the development of turbulence over the surface [35, 37]. The formulae from engineering texts must be used with caution, bearing in mind that the actual resistance is likely to be lower than the calculated value.

The tendency of natural convection to occur depends on the dimension of the leaf and the temperature difference between the hot or cold surface and the surrounding air. This tendency is measured by Gr, the Grashof number:

$$Gr = agd^3(T_s - T)/v^2 \qquad (3.6)$$

where a is the coefficient of expansion of the fluid, d is the characteristic dimension of the leaf, g is the acceleration due to gravity and $(T_s - T)$ is the temperature difference between surface and air. Inserting appropriate values of a and v for air at 20°C, Equation (3.6) becomes:

$$Gr = 158d^3(T_s - T) \qquad (3.7)$$

Monteith [38] suggests that when $Gr > 16 \, Re^2$ the predominant mode of transfer is natural convection, whereas, when $Gr < 0.1 \, Re^2$ it is forced convection. Hybrid convection is presumed to occur at intermediate values.

Equations for the calculation of boundary layer resistances are given in Box 3.1, and results of some calculations appear in Fig. 3.5.

3.2.2 Boundary layers over canopies

In field studies of exchange rates between plants and the atmosphere, the canopy as a whole is often taken as the active surface. Analysis of appropriate micrometeorological profiles over the canopy can provide good estimates of exchange rates. The approach first used to find carbon dioxide assimilation rates of cereal crops, has been extended to forests and natural vegetation (and, incidentally, used to measure deposition rates of atmospheric pollutants). The analysis provides some important insights into the significance of certain physiognomic attributes of vegetation, such as plant height and spacing.

Vegetation, because it is rough, slows down the air flow and creates a turbulent boundary layer. Transport of heat, water vapour and carbon dioxide through this boundary layer occurs by turbulent diffusion, at a rate determined by the turbulent structure of the air which, in turn, is determined by the wind speed and the aerodynamic roughness of the vegetation.

As the wind speed at the surface of all leaves is zero, but finite in the atmosphere over the vegetation, it follows that the plants are a *sink* for atmospheric momentum. In the daytime, the leaves are usually a sink for carbon dioxide and a *source* for water vapour and heat. These relationships are expressed as follows:

Flux of any entity $\qquad F = -K \times$ concentration gradient $\qquad (3.8)$

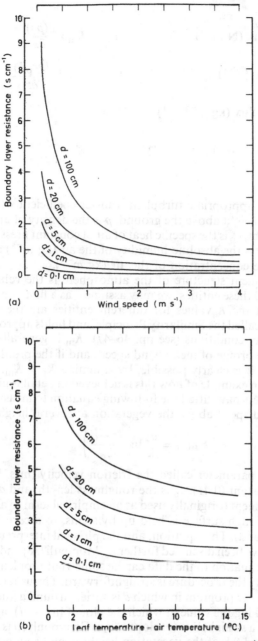

Fig. 3.5 Boundary layer resistances for water vapour flux from leaves with dimensions $d = 0.1$ cm − 100 cm. In (a) the resistance to forced convection as a function of wind speed. In (b) the resistance to free convection as a function of the difference in temperature between leaf and air. In both cases the resistances are calculated as shown in Box 3.1 for *one* surface of the leaf. Care must be taken in calculating a resistance to transfer processes at *both* surfaces when, by convention, using only the plan area (i.e. the area of one surface) as the basis for expressing leaf area (see Box 3.2).

Momentum flux (N m^{-2}) $\tau = -K_m \times \dfrac{\partial(\rho u)}{\partial z}$ (3.9)

Heat flux (J m^{-2} s^{-1}) $C = -K_H \times \dfrac{\partial(\rho c_p T)}{\partial z}$ (3.10)

Water vapour flux (kg m^{-2} s^{-1}) $E = -K_E \times \dfrac{\partial \chi}{\partial z}$ (3.11)

Carbon dioxide flux (kg m^{-2} s^{-1}) $P = -K_p \times \dfrac{\partial c}{\partial z}$ (3.12)

where K is the appropriate turbulent transfer coefficient which always increases with height above the ground, ρ is the density of air, u is wind speed, z is height, c_p is the specific heat of air at constant pressure, T is air temperature, χ is the absolute humidity of the air (kg m^{-3}) and c is the carbon dioxide concentration of the air (kg m^{-3}).

As the turbulent structure of the atmosphere is the vehicle for the transport of all these entities, we may assume – as a first approximation at least – that the K values for different entities are the same. This assumption is called the Similarity Principle, and holds approximately in neutral stability conditions (see pp. 46–47). K_m is generally found by analysis of the profile of mean wind speed, and if the profile of CO_2 is also recorded, it is clearly possible, by assuming $K_p = K_m$, to find the flux of CO_2. An example of how this is achieved in detail is given in [39].

Many workers have fitted the following equation to measurements of the mean wind speed above the vegetation at several heights (z):

$$u(z) = \frac{u_*}{k} \ln \left(\frac{z - d}{z_0} \right)$$ (3.13)

where u_* is a parameter called the friction velocity (m s^{-1}), k is von Karman's constant (0.41), z_0 is the roughness length, and d is the zero plane displacement (originally used as an empirical term to allow for the fact that the rough surface is held up by stems, to a certain height d, above the ground). The equation, and the physical interpretation of the parameters have been discussed further by Monteith [38] and Thom [40].

Fitting the equation to the data can be somewhat problematic, but in neutral stability the procedure is straightforward. The usual approach is to use a computer program in which d is varied around a guessed value, eventually picking the value at which the plot of $\ln (z - d)$, against wind speed u, yields a straight line (of slope k/u_*). Commonly, d is found to be between 0.6 and 0.8 of the vegetation height h, and so a good starting guess would be 0.7 h. To test for a straight line the correlation coefficient can be calculated after each trial. The intercept gives z_0, a very important attribute of the vegetation called the roughness length, which is a measure of the momentum absorbing power of the vegetation (Fig. 3.6).

For most crops it is found that $z_0 \simeq 0.1 \, h$, where h is the height of the

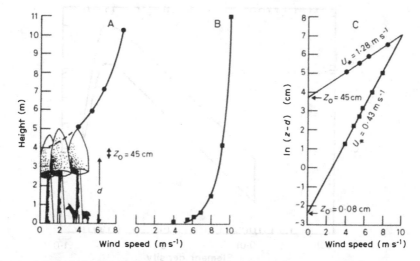

Fig. 3.6 Typical wind profiles over (A) tall vegetation and (B) short vegetation. A and B show raw data for tall and short vegetation, respectively, and C shows the logarithmic plot used to find the roughness length z_0 and the friction velocity u_* Note that the tall vegetation has a much higher value of z_0 and u_* than the short.

vegetation. In wild vegetation this relationship often holds, but may break down because of extreme values of plant spacement or a lack of homogeneity. If plants are very sparse or very crowded z_0 falls considerably below 0.1 h: data on this are given in Fig. 3.7.

Once the equation has been fitted, it is possible to find the turbulent transfer coefficient K_m. This is not a constant, but increases with height above the ground. At any height, its value can be found from the relationship derived in Monteith [38]:

$$K_m = ku_*(z - d)$$

It is shown in [38] that the resistance which we can think of as connecting the vegetation to the atmosphere, denoted as r_a^m and written with a subscript a and suffix m to show that it is the aerodynamic resistance appropriate to momentum transfer, is:

$$r_a^m = \left\{ \ln (z - d)/z_0 \right\}^2/(k^2 u(z)) \qquad (3.14)$$

or, once u_*^2 is known:

$$r_a^m = u(z)/u_*^2 \qquad (3.15)$$

Equation (3.14) may be used to investigate the effect of vegetation height on r_a^m: simply assume that $z_0 = 0.1\ h$ and $d = 0.7\ h$, and that z_0 and d are insensitive to wind speed (in practice they may vary a little due to the stems' bending and the plants' parting at higher wind speed). The result shows that resistances over short vegetation are much higher than those over tall vegetation at a given wind speed (Fig. 3.8).

45

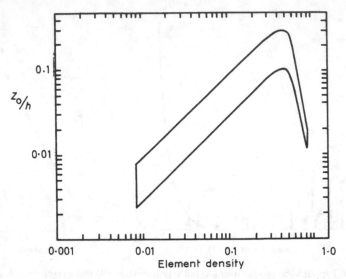

Fig. 3.7 The influence of vegetation density on the fraction z_0/h, where z_0 is the roughness length and h is plant height. Element density is defined as the plant silhouette area normal to the wind per unit ground area occupied by each plant. For most crops it can be assumed $z_0 = 0.1\ h$, but for sparser or denser vegetation this cannot be assumed. (From Garrett [41].)

Three further complications arise in the analysis of the wind profile:

(a) *Fetch* The wind profile must be measured in a fairly large area of homogenous vegetation if it is to be properly characteristic of that vegetation. As the air flows from one vegetation type to another, the profile 'adjusts' to the new roughness characteristics. The distance over the specified vegetation which the air has traversed is termed *fetch*. At low fetch, the logarithmic profile immediately above the vegetation can be expected to show an abrupt change of slope, the part of the profile immediately above the vegetation having a slope which is characteristic of the local roughness, whilst the upper part of the profile has a slope characteristic of the terrain over which the air previously flowed. As a rule of thumb, the fetch should be at least two hundred times the height of the anemometer mast. For miniature cup anemometers it is not practical to have a mast less than 1 m tall, so a fetch of 200 m represents a minimum.

(b) *Buoyancy* The analysis of the wind profile outlined above is appropriate only in meteorological conditions known as *neutral*, i.e. when the vertical gradient of air temperature is about $-0.01°C\ m^{-1}$. Neutrality occurs around sunrise and sunset and on dull days and cloudy nights. When the vegetation is most active, conditions are often *unstable* with temperature decreasing rapidly with height or *stable* with tempera-

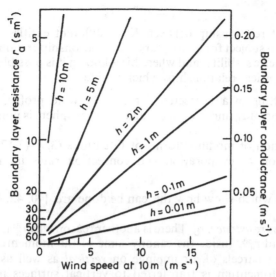

Fig. 3.8 The boundary layer resistance of vegetation as a function of the wind speed at 10 m above the ground. The calculation has been done for several values of vegetation height (h).

ture increasing with height. In the first case, there is a tendency for parcels of air to rise like bubbles, increasing vertical transport. In the second case, vertical motion of eddies is suppressed and transport is decreased. Both these effects distort the wind profile and, if not allowed for, prevent its analysis. Thus, methods have been devised to make a *stability correction* to the wind profile, based on an index of stability.

One index of stability is the Richardson number:

$$\text{Ri} = \frac{g}{T} \times \frac{(\partial T / \partial z)}{(\partial u / \partial z)^2} \tag{3.16}$$

Usually Ri is calculated from observations of temperature and wind speed at levels z_1 and z_2, by applying Equation (3.16) in finite difference form. Attention must be paid to the sign of $(\partial T / \partial z)$, which is negative when the temperature declines with height. In neutral conditions, where forced convection occurs, $\text{Ri} = 0 \pm 0.01$. Negative numbers outside this range denote unstable conditions with a contribution from free convection, whereas, positive numbers denote that forced convection is damped.

The similarity principle does not hold outside a limited range of Ri. In general, unstable conditions are those in which vertical heat transport is more effective than momentum transport:

$$K_H \simeq K_{gas} > K_m \tag{3.17}$$

and so:

47

$$r_a^H \simeq r_a^{gas} < r_a^m \qquad (3.18)$$

The exact relationship between K for different entities has been a controversial subject for many years. The relationship certainly depends on atmospheric stability, and when this is known it is possible to refer to existing empirical relationships which enable:

1. The plotting of a 'corrected' form of the wind profile, and hence evaluation of z_0 and d, even though the atmosphere is non-neutral in stability.
2. Application of modified forms of Equations (3.9)–(3.12), enabling calculation of evaporation and convection rates in non-neutral stability.

Key papers which show how this can be done are: [40, 42, 43, 44, 45].

(c) *Excess resistance*, r_b There is another reason why r_a^m is unlikely to equal r_a^H and r_a^{gas}. Inside the canopy some of the momentum transfer occurs when parcels of air impinge on *vertical* as well as *horizontal* surfaces. Momentum is transferred to vertical surfaces much more effectively than are mass and heat. Consequently, r_a^m is less than r_a^H and r_a^{gas}, the difference being termed excess resistance r_b. According to Thom [40], the magnitude of r_b is typically $4/u_*$.

3.2.3 Air within canopies

Attempts to quantify the relationship between wind speed and height within the canopy have not been particularly fruitful, though some progress has been made (see [46] and Fig. 3.9). It would be useful to be able to estimate the within-canopy profile from a knowledge of the distribution of leaves and the wind speed over the canopy. Such estimates would facilitate the calculation of the average r_a in successive layers of leaves. (It might also be valuable to animal ecologists who often want to know the microclimate for birds and other animals.)

Some authors have attempted to measure directly, the turbulent transfer coefficient, K_m. One method used was to release a marker gas (N_2O) from tubes on the ground and then measure its upward diffusion, using an infra-red gas analyser [47]. Druilhet *et al.* [48] made use of the minute natural emanations of the radioactive gas thoron, which come from the ground. Results show that profiles of K are very uneven, but the general relationship that might be postulated, is that the logarithm of K increases with height above the ground (Fig. 3.10). At the soil level, K is only an order of magnitude or so more than the coefficient for molecular diffusion (1.5×10^{-5} m^2 s^{-1}), whereas, at the top of the vegetation it assumes the sort of value estimated from ordinary analysis of the wind profile over the vegetation (Fig. 3.10).

Since the definition of the resistance to transfer between z_1 and z_2 is:

$$r_a^m(z_1, z_2) = \int_{z_1}^{z_2} \frac{\partial z}{K_m} \qquad (3.19)$$

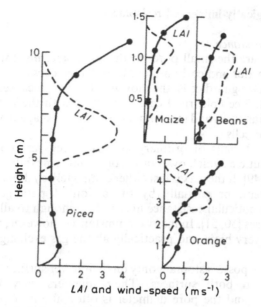

Fig. 3.9 Profiles of wind speed within crops (●) in relation to the leaf area index (---). (From Landsberg and James [46].)

it becomes possible to estimate values of the vertical resistors connecting layers of leaves. For example, if we suppose that near the middle of the canopy K is 10^{-3} m^2 s^{-1}, then 1 m of air would constitute a resistance of 1000 s m^{-1}, an extremely high value compared with the resistance over the canopy and very high in relation to the overall stomatal resistance of the canopy (Table 4.1, p. 63).

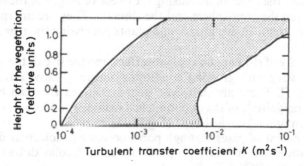

Fig. 3.10 Observed magnitude of the turbulent transfer coefficient, K, within vegetation. The hatched area encompasses the range of experimental values (see Legg [47] and Druilhet [48]). Note that the horizontal axis is logarithmic and that there is a marked trend towards lower values near the ground.

3.3 Physiologically-influenced resistances

3.3.1 Leaf resistance

The stomata are the small pores in the leaf epidermis, through which most of the water and carbon dioxide exchange occurs. Each stoma is bounded by two guard cells, the movement of which causes the pore to open or shut. The epidermal cells in contact with the guard cells are called subsidiary cells and may or may not differ in shape from the rest of the epidermal cells.

The external walls of ordinary epidermal cells are coated with a waterproof cuticle, which owes much of its impermeability to a surface layer of wax [49]. If the wax is lost either completely by dipping the leaf in organic solvent, or partially by the action of pathogens or wind abrasion, the cuticular resistance may be low enough to allow substantial water losses [50, 51]. In many circumstances, however, the cuticular resistance is very high, and practically all the gas exchange occurs via stomata.

Some leaves possess stomata only on their lower surface, while others have them on both surfaces. Their numbers vary from about 50–500 mm^{-2} and the pore diameter is often about 10 μm. Stomata occur on other plant parts such as petals and fruit. A theoretical relationship between number of pores, their size and the consequent diffusion resistance of the leaf is given by Rutter [52].

Moss gametophytes and algae do not possess stomata, although recent studies show that many mosses have well-developed cuticular wax, which may constitute a substantial cuticular resistance [53]. Brown algae, as their thalli dry out between tides, display appreciable surface resistances related to their ranking in position on the sea shore [54], though they do not possess stomata and the exact nature of this resistance is poorly understood.

In some species the stomata are sunk into pits, located in grooves in the leaves, surrounded by hairs, or protected by leaf rolling. All these features have the effect of increasing the effective length of the diffusion path between the atmosphere and the stomata. They are interpreted as adaptations to life in dry places, and plants possessing them are known as *xerophytes*.

The density of stomata on the leaf surface and the size of the stomatal apparatus vary with growing conditions and may show intraspecific variation [55]. There are, however, generalizations that can be made about the magnitude of the minimum leaf resistance and its relationship to life form and ecological type (Fig. 3.11).

As movement of gases through pores occurs by molecular diffusion the resistances are inversely proportional to molecular diffusion coefficients, i.e. at 20°C:

$$r_s^{H_2O}:r_s^{CO_2}:r_s^{SO_2} = 1:1.65:1.98$$

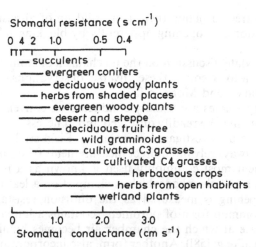

Fig. 3.11 Stomatal resistance and conductance – a survey of different ecological groups. (From Körner *et al.* [55].)

The effects of external and internal stimuli on the stomata have been investigated by botanists over a period of 300 years. Early observations showed that stomata open when the leaf is illuminated and usually shut at night; that they are subject to endogenous 'sleep' rhythms; that stomata shut when the plant suffers water stress; and that stomata open when the CO_2 supply is depleted, as when the underlying mesophyll is rapidly photosynthesizing, and close when the CO_2 supply is enhanced (see [56]). It was also realized that the opening movement of the guard cells was affected by an increase in pressure inside them caused by passage of water from the surrounding cells.

A satisfactory explanation of stomatal movement should take into account the mechanical properties of the stomatal apparatus, showing how a change in turgor pressure within the guard cells can account for opening movements; and it should show how the generation of turgor is linked *via* metabolism to environmental stimuli. At present, there is no consensus on these matters, although recent discoveries suggest that thorough understanding of the metabolic aspects may not be far away. The following observations seem critical:

1. Stomatal opening is much more sensitive to blue light than to red, and so not simply driven by photosynthesis.
2. When stomata open, K^+ ions enter guard cells from the surrounding tissues causing a decrease in osmotic potential which, in turn, causes water to enter so that the turgor pressure increases. The charge imbalance that would be caused by the K^+ flux is prevented by an associated movement of Cl^- or organic anions.
3. Epidermal tissues, and almost certainly guard cells themselves, are rich in PEP carboxylase and malic enzyme.

4. In water-stressed plants there is a build up of abscisic acid which prevents stomatal opening apparently by blocking K^+ uptake.

For an up-to-date discussion on the mechanism of stomatal movement which takes into account these diverse observations, the reader is referred to Jarvis and Morison [57].

Laboratory studies are usually conducted on convenient species in which stomata may be readily observed and, if necessary, the epidermis peeled away to be investigated in isolation. Field studies on the other hand nearly always rely on indirect measurements of stomatal opening. The instrument most commonly used is known as a porometer and consists of a small chamber which is clamped on a leaf or shoot. The degree of opening is measured as the diffusion resistance to water vapour. A common form of porometer contains a humidity sensor to record the rate at which the chamber air becomes humidified by the transpiring leaf (e.g. [58]). Another form, also incorporating a humidity sensor, depends on the measurement of the rate at which dry air must be bled into the chamber, to exactly balance the transpiration by the leaf, so that humidity is unchanged [59]. Most porometers are calibrated in appropriate units of resistance (s cm^{-1}), or conductance (cm s^{-1}).

To characterize the stomatal conductance of a stand of vegetation with a porometer requires fairly extensive recording, preferably at different levels and on different plants. A typical day's observations are shown in Fig. 3.12.

The environment may be changing so rapidly, in relation to the time needed for stomata to react, that the stomata cannot keep pace with it and never reach a steady aperture. This seems to be especially true for conifers in which stomata can take an hour or more to respond to a change in light levels. Variables are frequently correlated with each other, confusing interpretation of results. Their separate effects can be evaluated only in laboratory studies in which variables are altered one at a time. Conductance may be written as a function of each environmental variable (often on the basis of previous knowledge gained from laboratory studies) and then fitted to the observations by a suitable optimization procedure (see [61]).

Stomatal conductance in the field is often found to depend upon the following variables:

1. Photon flux density.
2. Leaf water potential.
3. Temperature.
4. Atmospheric humidity or the leaf to air vapour pressure difference.
5. Carbon dioxide concentration (although this often does not change much in the field).
6. Age of the leaves or season of the year.

Typical responses to these variables are shown in Fig. 3.13.

52

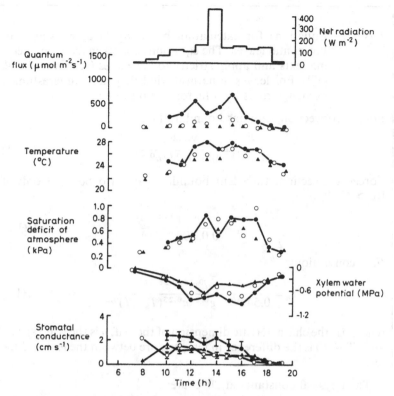

Fig. 3.12 Stomatal conductance and environmental variables in a tropical timber tree *Tectona grandis* (teak). Data collected during the wet season in Nigeria, with cloud cover for most of the day. Readings taken at three levels in the canopy: top (●), middle (○), bottom (▲). (From Grace, Okali and Fasehun [60].)

Stomata also respond to air pollutants, closing in response to photochemical smog and ozone, and opening when exposed to SO_2 in the range occurring in polluted air (50–1000 mm³ m⁻³).

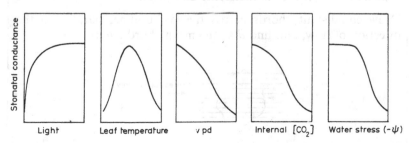

Fig. 3.13 Diagram to show typical response curves of stomatal conductance to five variables. (Adapted from Jarvis [61].)

Box 3.1 Equations for calculating boundary layer resistance of horizontal leaves. The equations are taken from engineering texts and apply strictly to flat plates (see Grace *et al.* [37]). For leaves in natural wind they may underestimate exchange rates by a factor of up to two.

Forced convection, laminar boundary layer:

$$r_a = \frac{d^{0.5}v^{0.17}}{0.66\,D^{0.67}u^{0.5}} \qquad (B1.1)$$

Forced convection, turbulent boundary layer (to be used only if Re $> 17\,000$):

$$r_a = \frac{d^{0.2}v^{0.25}}{0.03\,D^{0.67}u^{0.8}} \qquad (B1.2)$$

Free convection:

$$r_a = \frac{d^{0.25}v^{0.25}}{0.54D^{0.75}g^{0.25}a^{0.25}(T_s - T)^{0.25}} \qquad (B1.3)$$

where d is the characteristic dimension of the leaf, u is the wind speed, and $(T_s - T)$ is the difference in temperature between the leaf and the air.

The physical constants at 20°C are:

  a, coefficient of thermal expansion of air ($\approx 1/293$ K^{-1})
  D, diffusion coefficient in air (cm^2 s^{-1}), for heat (0.21), carbon dioxide (0.15) or water vapour (0.24)
  g, acceleration due to gravity (981 cm s^{-2})
  v, kinematic viscosity of dry air (0.15 cm^2 s^{-1})

Hybrid (free and forced) convection:

Evaluate (B1.1) and (B1.3), then combine resistances in parallel.

Calculation of d, the characteristic dimension:

Draw equidistant chords across the leaf outline, parallel to the direction of flow, and find d as the mean chord length.

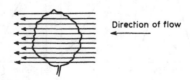

Box 3.2 Rules for applying resistance analogue to leaves, on a plan area basis.

Resistances with an arrow (↑ or ↓) are determined for each active surface separately. They may be combined to form resistances (without arrows) which are appropriate to the whole leaf and which when used in Equation (4.4) yield values of C and λE on the basis of plan leaf area. Subscripts s and a refer to stomata and boundary layer, superscripts H and H_2O refer to heat and water vapour.

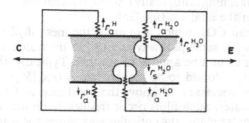

For water vapour

$$\frac{1}{(r_s^{H_2O}+r_a^{H_2O})}=\frac{1}{(\uparrow r_s^{H_2O}+\uparrow r_a^{H_2O})}+\frac{1}{(\downarrow r_s^{H_2O}+\downarrow r_a^{H_2O})} \quad (B2.1)$$

If $\uparrow r_s^{H_2O} \neq \downarrow r_s^{H_2O}$ or $\uparrow r_a^{H_2O} \neq \downarrow r_a^{H_2O}$ evaluate (B2.1) from separate determinations of resistances for adaxial and abaxial surfaces. If there are no stomata on one surface, set the appropriate r_s to infinity. If amphistomatous with aerodynamically similar surfaces then:

$$r_s^{H_2O} + r_a^{H_2O} = (\uparrow r_s^{H_2O} + \uparrow r_a^{H_2O})/2 \quad (B2.2)$$

For heat

$$\frac{1}{r_a^H} = \frac{1}{\uparrow r_a^H} + \frac{1}{\downarrow r_a^H} \quad (B2.3)$$

If

$$\uparrow r_a^H = \downarrow r_a^H$$

then

$$r_a^H = \uparrow r_a^H/2$$

3.3.2 Resistances involving CO_2

In the analysis of the diffusion pathway for water vapour, the resistances involved are just r_s and r_a, as the source of water is the air in the substomatal cavity which is saturated with water vapour, and the sink is the atmosphere. In the case of CO_2, there is an additional length of diffusion pathway between the stomata and the sites of carboxylation (Fig. 3.1). Transport of CO_2 in this region is not exclusively in the gas phase and not necessarily strictly diffusional. Gaastra [62] used the term mesophyll resistance (r_m) to describe the apparent resistance in this region. Its value is found from comparison of simultaneous measurements of the rates of transpiration and photosynthesis. Today, the term mesophyll resistance has been replaced by *residual resistance* to imply that the physical meaning is not really known, but that the value arises as a residual term in the analysis (see [36]).

Gases other than CO_2, which also enter the mesophyll tissues, are unlikely to follow the same pathway as CO_2, so the residual resistance found for one gas cannot be applied to another. Typically, the residual resistance for CO_2 is found to be higher than r_s (see [55]).

It is sometimes necessary to know the resistance to CO_2 transfer across a film of water. Such films occur frequently on mosses, lichens and algae. In vascular plants the cuticular wax causes the water to form droplets that run off, and the lower surfaces of leaves are, of course, sheltered from the rain. As the diffusion coefficient of CO_2 in water is about 0.16×10^{-4} cm^2 s^{-1}, compared to that in air of 0.16 cm^2 s^{-1}, it follows that even a thin film constitutes a substantial resistance and is likely to limit the rate of photosynthesis in bright light.

3.4 Micrometeorological stomatal resistance

The apparent stomatal resistance of the canopy as a whole can be found from suitable micrometeorological measurements, as an alternative to the task of measuring stomatal resistance with a porometer. Important papers showing the development of this approach are Cowan [22], Thom [40], Stewart and Thom [63] and Monteith [64].

4 Heat and water exchange at plant surfaces

For where-e'er the sun does shine,
And where-e'er the rain does fall,
Babe can never hunger there,
Nor poverty the mind appall.
William Blake

4.1 Introduction

A knowledge of the resistances to diffusion, as discussed in the previous chapter, may be applied to answer some important questions about the water and heat relations of plants:

1. How does the temperature of leaves, buds or stems depend on atmospheric and plant variables?
2. How much water does a plant use in a specified environment and how does this depend on changes in (a) atmospheric variables and (b) the physiological state of the plant?

Answers to these questions are of concern in practical fields such as agriculture and hydrology. Moreover, they shed light on problems of plant adaptation to the extremes of desert and arctic environments.

Transport of heat and water vapour from the leaf surface to the atmosphere occurs through a common boundary layer, at least in an amphistomatous leaf. The relationship between these entities is further linked by the strong effect of temperature on the concentration of water vapour in the water-saturated atmosphere of the substomatal cavity. If the leaf is cooled, this concentration will decline and so the driving gradient for transpiration will be diminished. Thus the leaf will transpire more slowly. If the leaf is heated it will transpire at a faster rate (all other variables being kept constant). The magnitude of this effect, over ordinary temperature ranges, is considerable: air at 22°C holds twice as much water vapour as air at 10°C (see Appendix 5). Transpiration is therefore closely linked to leaf temperature.

4.2 Energy balance

Having shown in general terms how transport of heat and water vapour are related, we can now write equations to express the relationship in quantitative terms.

The energy balance of the leaf in bright sunlight can be written as:

Heat gains by radiation = Heat losses

$$\mathbf{R} = \lambda \mathbf{E} + \mathbf{C} + \mathbf{S} + \mathbf{G} + \mathbf{P} \qquad (4.1)$$

where **R** is the net heat gained through radiative exchanges, allowing for the absorption of both downward and upward components of long and shortwave radiation, **E** is the evaporation rate multiplied by the latent

57

heat of vaporization (λ) in order to express evaporation rate in units of energy, C is the heat lost by convection, S is the rate at which heat goes into storage within the leaf, G is conduction of heat down the petiole, and P is the rate at which energy is being trapped in chemical bonds by photosynthesis. All the terms are in appropriate units of energy flux, preferably $W\ m^{-2}$.

This can be generalized to cover any time of day by denoting gains of energy as positive and losses as negative. Then, since energy is neither created nor destroyed, the algebraic sum of the energy fluxes must be zero:

$$R + \lambda E + C + S + G + P = 0 \tag{4.2}$$

In practice G and P are quantitatively unimportant, except in a few special cases, and S is not important if we are thinking of the energy balance averaged over minutes, as opposed to seconds. Dropping G, P and S, we can say that the net energy absorbed is partitioned between latent heat (λE) and sensible heat (C):

$$R = \lambda E + C \tag{4.3}$$

Both λE and C can be written in terms of variables which, at least in principle, can be measured. Using an Ohm's law analogy they can be expressed in terms of a driving gradient divided by an appropriate resistance:

$$R = \frac{\rho c_p (e_s(T_s) - e)}{\gamma(r_s + r_a^{H_2O})} + \frac{\rho c_p (T_s - T_a)}{r_a^H} \tag{4.4}$$

where ρc_p is the volumetric heat capacity of the air ($J\ m^{-3}\ {}^\circ C^{-1}$) made up of the density (ρ) and the specific heat at constant pressure (c_p), $(e_s(T_s))$ is the saturated vapour pressure of air at the temperature T_s of the leaf surface, e is the water vapour pressure of free air (all vapour pressures in mbar), γ is the psychrometric constant (0.66 mbar $^\circ C^{-1}$), and r_s, $r_a^{H_2O}$ and r_a^H are stomatal and the boundary layer resistances ($s\ m^{-1}$).

In practice, considerable care is required in applying these equations to leaves, as different authors have used different conventions in the definition of leaf area, and as leaves may be amphistomatous (stomatal on both sides) or hypostomatous (stomata restricted to the underside of the leaf). Moreover, the term R is the net heat gain, obtained by adding the net downward fluxes (all-wave), to the net upward fluxes, and cannot be equated with net radiation as conventionally measured, which is the downward minus the upward flux ($W\ m^{-2}$).

In applying measured or calculated resistances, it is advisable to treat the two surfaces of the leaf as two sides joined in parallel (Box 3.2) to give values of C and λE on a plan area basis. The radiation absorbed R must also be expressed on a plan area basis if it is to be equated to $(C + \lambda E)$ in Equation (4.3).

58

4.3 Calculation based on the heat balance

4.3.1 Leaf temperature and transpiration

Equation (4.4) can be used to estimate the temperature that a specified leaf would attain if subjected to any particular environment. The leaf is specified by its stomatal and boundary layer resistance, although the latter also depends on wind speed (Box 3.1). It is instructive to solve Equation (4.4) with a computer, using an iterative procedure, when the process of balancing the equation becomes analogous to the physical process whereby the leaf eventually comes to a steady temperature. The sequence of calculation is:

1. Select required values for the parameters R, e and T_a, which define the physical environment.
2. Select appropriate values for r_s and r_a, the latter taking into account wind speed and leaf dimension. These define the leaf.
3. Increment T_s, starting at some low temperature (say 0°C) and rising in one degree steps. At each step the leaf to air vapour pressure difference $(e_s(T_s) - e)$ increases, and thus, the evaporation term increases. The temperature difference $(T_s - T_a)$ also increases, though at a different rate, so convective heat loss increases. The physical 'constants' do vary a little with temperature, and this should be incorporated into the computer program.
4. Find the temperature T_s at which $(\lambda E + C) \simeq R$. If necessary, a more accurate run can then be done using smaller increments in a narrower range, eventually converging on the exact values of T_s, λE and C (steps as small as 0.1°C are recommended). Then T_s and λE may be plotted as a function of R, e, T_a, r_s or r_a (Figs 4.1 and 4.2).

Note the analogy between the computation and the physical process: a leaf absorbing energy at a certain rate R, will gradually heat up until evaporation and convection attain rates which, when added together, exactly balance the rate of absorption of radiant energy.

The following general conclusions can be drawn from the analysis (Figs 4.1 and 4.2):

1. Transpiration is strongly coupled to radiation absorption, particularly when r_a is large, as in the case of large leaves at low wind speeds.
2. An increase in r_s is always accompanied by some decline in the transpiration rate, though the effect is much greater when r_a is small, as in needle-like leaves or when the wind speed is high.
3. Wind speed influences transpiration in a complex manner through r_a. At high rates of energy absorption an increase in wind speed (i.e. a decrease in r_a) causes a decline in surface temperature and thus, transpiration rate is substantially reduced. This is contrary to general belief (which, incidentally, is much influenced by casual observations of wet garments on clothes lines where $r_s \simeq 0$, until the surface dries

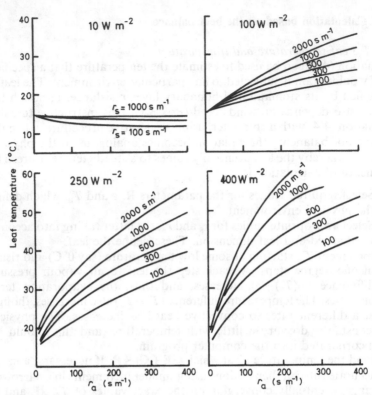

Fig. 4.1 Leaf temperature, calculated as a function of the boundary layer resistance, stomatal resistance and available energy. In all calculations the air temperature was 15°C and the water vapour pressure was 9 mbar. Refer to Fig. 3.5 to obtain r_a from a knowledge of leaf size and wind speed.

out), but the phenomenon can easily be shown experimentally (e.g. [65]). However, at low rates of energy absorption (e.g. **R** = 10 W m^{-2}), the leaf is cooler than the air and then an increase in wind speed does increase the transpiration rate.

4. In some cases, the transpiration rate in W m^{-2} exceeds the net supply of radiant energy, **R**. In these circumstances the leaf is cooler than the surrounding air and so the flow of sensible heat is from the air to the leaf (i.e. the sign of **C** is reversed). An extreme case of this is the wet bulb of an aspirated psychrometer in which $r_s \simeq 0$ and r_a is also very low.

5. Leaf temperatures may greatly exceed air temperatures, especially when the leaves are large. There is no physical limit of leaf temperature, as some authors have suggested. Large leaves are likely to be at a great disadvantage in hot dry places, especially where they are subjected to an additional component of **R** by reflection from bare ground, as in the desert. In this case the net energy load may well exceed the 400 W m^{-2} taken as the upper limit in Fig. 4.1, and

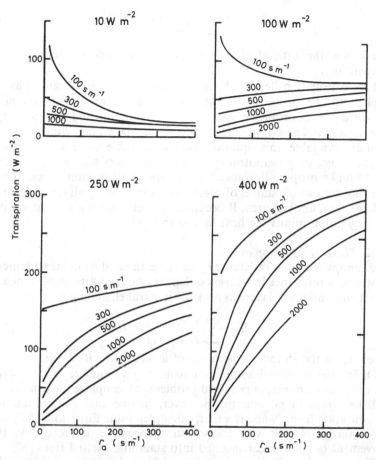

Fig. 4.2 Transpiration rate, calculated as a function of the boundary layer resistance, stomatal resistance and available energy. All other conditions are as for Fig. 4.1, with which this figure should be compared. Note that the transpiration rate is expressed as an energy flux, but can be converted to $g\ s^{-1}\ m^{-2}$ by dividing by the latent heat of vaporization of water ($2450\ J\ g^{-1}$ at $20°C$).

transpiration cooling is then likely to be crucial to the survival of the leaf. On the other hand, small leaves display a very low boundary layer resistance, never get much warmer than air temperature, so in hot dry conditions they are unlikely to be killed by high temperatures.

In fact, the iterative process described above is not the most efficient means of calculating transpiration. It can be shown that the following formula can be obtained from Equation (4.4), which enables transpiration to be calculated even though the surface temperature is not known [38, 66]:

$$\lambda \mathbf{E} = \frac{s\,\mathbf{R} + \rho c_p(e_s(T_a) - e)/r_a^H}{s + \gamma(r_a^{H_2O} + r_{st})/r_a^H} \quad (4.5)$$

where s is the rate of change of saturation vapour pressure with temperature.

This is usually called the Penman–Monteith equation, and may be applied to the problem of transpiration of leaves or of canopies, provided the appropriate values of the stomatal and aerodynamic resistances are known. In parts of the world where good climatological data are available, this equation has considerable value as a practical means of assessing evapotranspiration, as it is possible, through r_a and r_{st}, to make proper allowance for the type of vegetation cover. When working with whole stands of vegetation, and defining all terms relative to land area, not leaf area, $\mathbf{R}$ becomes the net radiation measured over the vegetation minus the heat flux to the soil.

4.3.2 The case of canopies
The canopy stomatal resistance r_{st} can be estimated from measurement of stomatal resistance in n layers of leaves using a diffusion porometer, and then treating the layers as resistors in parallel, finding r_{st} as follows:

$$\frac{1}{r_{st}} = \sum_{i=1}^{i=n} \frac{1}{r_{s,i}} \quad (4.6)$$

where $r_{s,i}$ is the stomatal resistance of leaves in the ith layer.

Difficulties in using this approach to finding stomatal resistance of the entire canopy arise from practical problems of sampling the canopy, as well as errors in porometry. However, in one case the cumulative transpiration from *Pinus sylvestris*, calculated using Equations (4.5) and (4.6), agreed closely with transpiration measured by recording the movement of radiotracer inserted into standing, excised trees [67].

The terms r_a^H and $r_a^{H_2O}$ in Equation (4.5) are best estimated from the wind profile, as outlined in the previous chapter, on the assumption that the source for water vapour coincides with the sink for momentum and that the Similarity Principle holds. In some cases, discussed by Thom [40], this simplification would lead to serious error, and modifications are required.

Canopy stomatal resistances reflect basic differences in the stomatal resistance of leaves, and in the amount of leaf per unit area of ground (Table 4.1). Herbaceous crops tend to have the lowest canopy stomatal resistances, though recent measurements on tropical tree plantations suggest that, in the favourable circumstances of the tropics, forests may also display low stomatal resistances. Otherwise, in the temperate world, forests generally have relatively high canopy stomatal resistances. The lowest boundary layer resistances occur in canopies of tall vegetation with small leaves, as already discussed in the previous chapter.

Applying the Penman–Monteith equation to the canopy as a whole and setting r_s and r_a to cover the range normally observed (Table 4.1), it

Table 4.1 Stomatal (r_{st}) and boundary layer (r_a) resistances for whole canopies

Reference	Canopy	r_{st} $(s\ m^{-1})$	r_a $(s\ m^{-1})$
Jarvis [59]	Grassland/heathland	50	20–50
	Agricultural crops	20	20–50
	Coniferous plantation	50	3–10
Cernusca and	Alpine sedge mat	63	12
Seeber [68]	Alpine pasture	56	15
Proctor [53]	*Mnium hornum*		
	cushion (Bryophyta)	30	100 at 0.3 m s^{-1}
Whitehead *et al.* [69],	Tropical tree		
Grace *et al.* [60]	plantation	16	70

is possible to make statements about the absolute amounts of water that would be used in specified conditions. This is important information to hydrologists who may wish to forecast the water yield of a proposed catchment, or to assess the change in catchment yield that might result from clearing a forest. For hydrological purposes, evaporation is generally calculated in mm of water for comparison with rainfall which is, by tradition, measured as depth of water in a rain gauge (Fig. 4.3).

The following conclusions can be drawn from Table 4.1 and Fig. 4.3:

1. The use of water by tall vegetation with small leaves (e.g. coniferous plantation forest) is, to a large extent, under the control of the stomata. Even when the energy absorbed is low, water use is likely to be closely linked to stomatal movements. Transpiration in this case is hardly sensitive at all to the absorption of radiant energy.

2. Minimal stomatal resistances of agricultural crops are generally quite low (Table 4.1). However, for much of the year, the crop is either absent from the site or in an early stage of development, when the leaf area index is low. Hence, average stomatal resistances are likely to be considerably higher than the value in Table 4.1, probably at least as high as for plantation forest. Thus, water use by agricultural crops, averaged over the whole season, may be lower than that of forests; and, since r_a is substantial, may not be so sensitive to stomatal movements.

3. Permanent, short vegetation like heathland and rough grassland, appears to display a higher stomatal resistance than agricultural crops, and a boundary layer resistance that is not very different from the latter (Table 4.1). The very low r_a recorded in alpine pasture is probably the result of very high wind speeds.

4. The available information for tropical trees suggests that they may be characterized by very low stomatal resistance and, because they often possess large leaves, a high boundary layer resistance. Water use in these conditions is relatively insensitive to small variations in

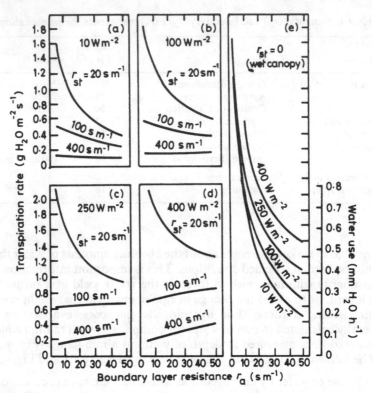

Fig. 4.3 Calculated transpiration rates from vegetation at four levels of available energy (10, 100, 250 and 400 W m^{-2}) and at three values of canopy stomatal resistance r_{st}. For comparison, (e) shows the rates of evaporation from wet canopy, calculated by assuming $r_{st} = 0$. In all cases, air temperature is 15°C and atmospheric saturation deficit is 8 mbar. The axis on the right hand side shows the transpiration and evaporation rates in alternative units, more familiar to hydrologists.

stomatal resistance and much more dependent on the absorption of radiant energy.

5. When the canopy is wet, the evaporation is usually estimated by assuming a film of water covers the leaf and so setting $r_s = 0$. Extremely high evaporation rates then occur from canopies with low boundary layer resistances (Fig. 4.3e), even at low levels of available energy. In these conditions, evaporation rate is much increased by wind and the surrounding air is cooled.

6. In most conditions, apart from those mentioned in (5), water loss is not very sensitive to moderate changes in wind speed.

It must be realized that the type of analysis presented here does not take into account the influence of the physical environment on r_{st}, such as the increase in resistance often caused by low humidity [70], nor does it take into account the feedback effect of the vegetation on the atmosphere. For example, one consequence of a high rate of transpiration in a large stand of vegetation would be an increase in humidity and a decrease in the temperature, which would in turn affect transpiration.

Precise estimation of the amount of water used by a particular type of vegetation would clearly require knowledge of the seasonal change in the bulk stomatal resistance. In agricultural crops it might be possible to estimate r_{st} from a knowledge of the seasonal pattern of leaf area index. An example of the seasonal pattern in canopy resistance for an agricultural crop in Britain is provided by Russell [71].

For hydrological purposes, it should be realized that transpiration rate is only one component of the hydrological cycle (see [52]).

The Penman–Monteith equation may also be applied to individual leaves when the symbols are referred to leaf area as originally in Equation (4.5) and not to land area, so that $\mathbf{R}$ is the net energy absorbed by the leaf. Note the presence of $e_s(T_a)$ again, which avoids concern with the temperature of the leaf itself. In order to apply the necessary weighting to upper and lower surfaces, the factor n is set at 1 or 2 for the amphistomatous or hypostomatous condition, respectively:

$$\lambda \mathbf{E} = \frac{s\mathbf{R} + \rho c_p (e_s(T_a) - e)/r_a^H}{s + n\gamma(r_a^{H_2O} + r_s^{H_2O})/r_a^H} \qquad (4.7)$$

When E is evaluated by this method the results are similar to those obtained by the iterative solution of Equation (4.4), although, of course, the latter provides an estimate of T_s and $\mathbf{C}$ as well as $\lambda \mathbf{E}$.

5 Field observations

*The science of vegetation is the study of the
morphology of plant communities.*
H. Meusel, translated in Richards [72]

5.1 Structure of vegetation

Botanical ecologists have long been concerned with the structure of
vegetation. The term structure is ill-defined but embraces floristic
composition, and includes the physiognomy or general appearance.

Based mainly on the height of the perennating buds above the ground,
Raunkiaer [73, 74] attempted to describe vegetation in terms of its
constituent *life forms* (Fig. 5.1). He pointed out that any stand of
vegetation contains a spectrum of life forms, which is broad in tropical
forest but restricted in climatic extremes such as on mountains, in high
latitudes or in deserts. In these latter regions, dwarf life forms
predominate. The notion of life forms is still useful today, and has been
used to draw attention to certain differences between sites which may be
associated with either climatic stress or grazing (e.g. [75, 76]). A similar
concept has recently been advanced by Lambert [77] in relation to
botanical surveying, though she speaks not of life forms but of 'eco-
organs'.

Raunkiaer also proposed a scheme of classification of leaves, based on
size (Table 5.1). The scheme was advocated by Richards, Tansley and
Watt [78] in a paper to the Imperial Forestry Institute, and used by some
forest botanists, especially those working in the tropics. According to
Richards *et al.* [78] 'Leaf size is often an important index of habitat' and

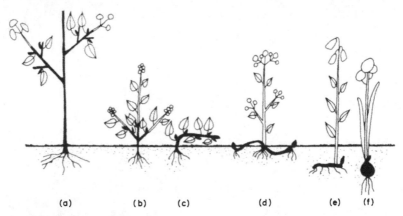

Fig. 5.1 Life forms defined according to relative positions of perennating parts. (a)
Phanerophytes, (b) and (c) Chamaephytes, (d) Hemicryptophytes, (e) and (f)
Cryptophytes. (Adapted from Raunkiaer [74].)

Table 5.1 Classification of leaves by size according to Raunkiaer, and the corresponding value of the characteristic dimension, d

Class	Area (cm^2)	Mean d (cm)
Leptophyll	0.25	0.35
Nanophyll	0.25–2.25	1.11
Microphyll	2.25–20.25	4.74
Mesophyll	20.25–182.25	10.06
Macrophyll	182.25–1640.25	30.2
Megaphyll	1640.25	40

strict adherence to Raunkiaer's scheme was recommended. Interest in leaf characteristics has been stimulated by palaeobotanical discoveries since the report by Bailey and Sinnot [79] concerning the correlations between the type of leaf margins in dicotyledonous plant and their distribution between climatic zones.

In the last two decades the application of physical principles to the study of vegetation has enabled a re-evaluation of structural attributes such as height and the size of the leaves. In this concluding chapter we will examine relationships between structure, function and environment, in the light of what has been said in preceding chapters.

5.2 Vegetation height
In Chapter 3 it was shown that boundary layer resistance depends on the height of the vegetation. Tall vegetation is closely coupled with the atmosphere, and so the temperature of the leaves and the air spaces between the leaves is unlikely to much exceed that of the atmosphere above. Hence, in a climate that is generally cold, short stature may be expected to confer selective advantage. There is much observational evidence to support this contention.

Achillea lanulosa in Sierra Nevada is a dwarf plant at the highest altitudes, becoming taller as sea level is approached. When individuals were taken from a range of altitudes and grown for several generations in a uniform garden, those from the higher altitudes remained dwarf (Fig. 5.2). Hence, the short stature of the high altitude populations in the

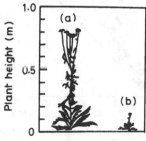

Fig. 5.2 Low and high altitude forms of *Achillea lanulosa*. (a) Collected at 1200 m above sea level. (b) Collected at 3300 m above sea level. (Redrawn from Clausen *et al.* [80].)

wild cannot simply be attributed to poor conditions for growth, but is an attribute which has arisen over many generations by a process of natural selection. Although the authors did not investigate the heat balance of different forms, and although several other causes of dwarfness cannot be ruled out, it seems likely that the main selection pressure in the mountain habitat has been the overall low air temperatures, in which short variants would have been favoured because of their higher leaf temperatures. Similar studies, in which it has been shown that high altitude populations are made up of prostrate genotypes, have been carried out on other species (e.g. [81, 82]).

There are also studies of the temperature regime of tall and short plants. Salisbury and Spomer [83] attached thermocouples to the leaves of cushion plants and erect plants growing at 3800 m in Colorado. The leaves of the cushion plants were found, on average, to be 5°C warmer when the sun was shining (Fig. 5.3).

In the case of trees near the altitudinal limit for tree growth, individuals become progressively more dwarfed as the tree-line is approached (e.g. [84, 85]). The extreme is a short and much contorted growth form, referred to as *krummholz*, in which the buds are within a metre or so of the ground (Fig. 5.4). Generally, krummholz vegetation gives way to dwarf shrubs at a higher altitude, which in their turn give way to completely prostrate vegetation. Proof that krummholz trees are genetic dwarfs is difficult to obtain, as they are not readily transplantable and vegetative propagules do not usually succeed, but it is almost certain that the dwarf form is indeed inherited in trees as well as in herbs [84].

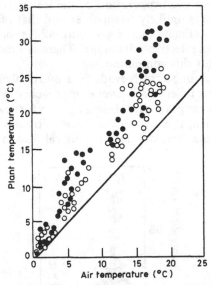

Fig. 5.3 Plant temperature at 3800 m in Colorado, classified according to whether the plant was erect (○) or cushion-like (●). The line shows the 1:1 relationship. (From Salisbury and Spomer [83].)

Fig. 5.4 *Pinus sylvestris* at the tree-line, Creag Fhiachlach, Cairngorm Mountains, Scotland. (a) *Pinus sylvestris* forms tall trees on the steep slope at 450 m above sea level. At higher altitudes (550 m) the trees are sparser and never tall. Some have dead limbs and are flagged (b). Others are shorter with twisted stems and closely-spaced growing points (c). Yet others are shorter still with prostrate stems (d). The main constituents of the vegetation at higher altitudes are dwarf woody plants, especially *Calluna vulgaris* and *Juniperus communis*. (Photos: M. Dixon and S. Allen.)

Prostrate growth forms also occur in plantations at high altitude. These are obviously not genetic dwarfs, and their condition may be caused by various types of damage including desiccation of leaves, breakage of branches by wind or snow, and frost.

According to Tranquillini [13] the main factors which lead to injury at the tree-line are frost and desiccation. The latter is related to temperature, as with low mean temperatures and a short growing season the phenological cycle is not properly accommodated and the cuticle is incompletely developed, causing an elevated cuticular conductance to water vapour. Whatever the mechanism, it seems that the climatic limit of tree growth is related to summer temperature: a survey of tree-line climates shows that they differ widely in mean winter temperature but they have in common a July mean air temperature of 10°C [16].

A distinction should be drawn here between the aerodynamics of plants that grow as scattered individuals and those that form major constituents of whole stands of vegetation. In the case of *Achillea lanulosa* and the plants studied by Salisbury and Spomer [83], the plants were distributed more or less as individuals. In such a case, the aerodynamic unit comprises the plants and their rocky terrain, rather than the vegetation stand. In this case, short stature is interpreted as short in relation to the roughness of the terrain. There are however, plenty of studies concerning extensive vegetation which may also be used to illustrate the general argument that short vegetation is warmer, and therefore fitter, in a cold environment.

Cernusca and Seeber [68] report temperature profiles above and within various alpine communities. It is useful to compare their profiles with those obtained elsewhere for tall vegetation (Fig. 5.5). In alpine pasture the temperature at the soil surface, roughly corresponding to the location of the meristems of the graminaceous constituents, was 25°C higher than that of the air above the canopy. In contrast, in tall forests, the temperature gradients are typically very small. In the example chosen for Fig. 5.5, the measured air temperature where leaves were most dense was less than 3°C higher than that above the canopy. Similar comparisons may be made with other variables, such as water vapour

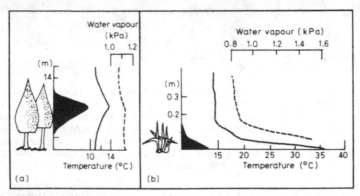

Fig. 5.5 Temperatures (solid lines) and water vapour pressures (broken lines) in canopies of (a) *Picea sitchensis* and (b) *Nardus stricta*. Note that within the forest canopy the gradients are never steep and the absolute values of temperature never very different from those in the atmosphere above. (From Jarvis *et al.* [86] and Cernusa and Seeber [68].)

pressure. Both sets of profiles in Fig. 5.5 were recorded near noon on a bright day in summer, and the ambient physical conditions were rather similar.

In terms of plant response, it should be realized that virtually all processes in the plant are enzyme controlled, and thus, sensitive to temperature. In particular, the temperature of the meristems may be more important than that of the leaves in determining how rapidly the plant develops. This is known to be true in grasses (e.g. [87, 88]) and if universally true, the definition of life forms based on location of the perennating buds in relation to the microclimatological profile, seems especially apt.

There are other consequences of tall stature apart from those related to heat exchange. Aerosols, such as fog and salt spray, are also transported to vegetation by turbulent transfer. In Japan, the trapping of small particles by coastal vegetation, has been investigated in connection with 'fog-preventing forests' – plantations intended to remove sea fogs that enshroud the land. Ingenious methods were employed to weigh the fog captured by the tree canopy, and to make comparisons with nearby grassy vegetation. In one test, the forests captured 33 mg m^{-2} s^{-1}, whilst the grass captured only a few tenths of this value [89]. Rutter [52] refers to examples in which rain gauges situated on the forest floor capture far more rain than equivalent rain gauges in the open, the difference being termed 'occult precipitation' and arising from interception of fog particles which coalesce and eventually drip to the forest floor.

Sea spray represents an environmental stress inasmuch as Cl$^-$ ions are toxic to plants, and sea salt, either on leaves or in the soil, contributes to a gradient of water potential and so may cause water stress. Moreover, tall vegetation near the sea frequently exhibits browned foliage after a storm. Boyce [90] gives a detailed account of species zonation at coastal areas of North Carolina (latitude 34°N). The vegetation displays a zonation which is equivalent to that evident at the tree line, with a herbaceous zone, a shrub zone and an arborous zone. A plausable hypothesis to explain this zonation is that tall vegetation, being so efficient at capturing particles including salt spray, cannot exist very close to the sea as it would experience a lethal dose of salt. Only short vegetation, which captures little salt, can tolerate proximity to the sea.

5.3 Leaf survival and design
The calculations of leaf temperature presented in the last chapter, raise important questions regarding plant survival and adaptive evolution. The calculations suggest that for leaf sizes ranging from $d = 10$ cm upwards, very high leaf temperatures may occur, and that transpiration must be an essential factor in preventing death by overheating.

Levitt [91] reviewed the rather extensive older literature on the occurrence of heat injury in nature. Many observations of 'burning' or 'heat injury', especially in horticultural material, may not have been

71

caused by high temperatures themselves, but by water stress occurring in hot dry periods of weather and mistaken for heat injury even though lethal temperatures were not reached. It is impossible to ascribe damage observed in the field to heat injury, unless careful studies are made of leaf temperature, coupled with laboratory experiments in which high temperatures are applied in conditions in which water stress is excluded. Such laboratory studies, in which exposure to a high temperature varied in duration from a few minutes to a few hours, suggest that for most ordinary plants the lethal temperature is about 50°C. There are differences in the lethal temperature according to the habitat in which the plant naturally grows: *Opuntia* and other succulents of the desert tolerate a somewhat higher temperature than most plants.

Leaf temperatures, actually recorded in plants in the wild, have often approached the known or presumed lethal point. Studies over many years by Lange have shown that this is true both for desert plants and for certain species in temperate climates [92, 93].

The desert cucurbit *Citrullus colocynthis* has rather large leaves for a desert plant, but remains below air temperature as a result of a high rate of transpiration [94]. When a leaf was excised to cut off its water supply, its temperature rose beyond the known lethal point, to 60°C (Fig. 5.6). It is well known that many desert plants display high rates of transpiration, their roots tapping a relatively large volume of the soil. Such plants have been termed 'water spenders': it is by spending water that they survive.

Very high tissue temperatures are not confined to desert plants, but have often been recorded in temperate plants and mountain plants growing in sheltered places. Larcher [95] found that the temperature in the centre of the alpine *Sempervivum montanum* rose to 54°C, or 32°C above that of the air (Fig. 5.7). This temperature was only slightly below the lethal point, at least in those parts of the plant closest to the ground. Other examples of plant temperature recorded in the field are given by

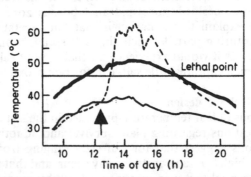

Fig. 5.6 Transpirational cooling in the desert plant *Citrullus*. Graph shows air temperature (heavy line), leaf temperature (thin line) and temperature of another leaf which was excised to stop transpiration (broken line). The arrow shows when excision was carried out. (From Lange [94].)

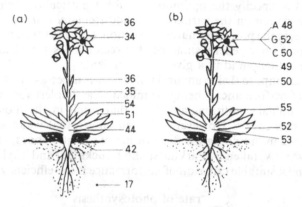

Fig. 5.7 Plant temperatures and lethal temperatures in *Sempervivum montanum*. (a) Plant temperatures on a sunny day in the Alps, (b) temperatures required to kill the tissue at various parts of the plant. A, anthers; G, gynoecium; C, corolla. (Adapted from Larcher [95].)

Gates [1], including measurements of his own which suggest some of the mechanisms by which leaf temperature is kept below the lethal point: from observations near Sydney, Australia, a contrast is shown between *Erythrina indica* in which temperature remained low as a result of the vertical posture of leaves; *Populus deltoides* which remained cool as a result of transpirational cooling; and *Jacaranda acutifolia* whose leaves were small and whose temperatures were thus closely coupled to that of air. Other useful observations on leaf temperatures of plants of diverse ecological type are reported by Stoutjesdijk [96].

Recent work has been concerned with the impact of high leaf temperatures on the photosynthetic performance of desert plants (e.g. Lange *et al.* [97]) and on the integrity of the photosynthetic apparatus [30]. The desert C4 plant *Tidestromia oblongifolia*, when grown in 'hot' regimes, displayed an optimum temperature for photosynthesis of 42°C, though thereafter the rate declined abruptly [30]. By contrast, the temperature optimum in the coastal C3 annual *Atriplex glabriuscula*, was not much affected by the growing conditions, and was around 30°C. The temperature at which irreversible inactivation of the partial processes of photosynthesis occurred suggested that activities residing in the chloroplast membranes were especially prone to heat inactivation: as temperature was increased it was the light harvesting system that broke down first, not the carboxylating enzymes.

The solution of the energy balance equation for leaves of specified size and resistance suggests a possible approach to the problem of leaf design: given basic information on the upper and lower lethal temperatures and knowing the response of photosynthesis to temperature and to changes in the component resistances of the diffusion pathway, it should be possible to specify the 'climate-space' within which leaves of a

73

particular design are able to survive. Going even further, some workers have tried to specify the optimum leaf for a particular environment, basing the design on the variables *size* and *stomatal resistance* [98, 99, 100]. Since, in the process of natural selection, only the fittest survive, we might expect some general similarity between the designed leaf and the leaves we actually find in a given environment.

In any attempt to define an 'optimal leaf' it is necessary to state what criterion of performance is being examined. The simplest view would be that the optimal conditions are those that result in the most photosynthetic production. However, it is well known that water stress is a major factor in moulding the flora of many climatic zones, and so an alternative view, taken by Parkhurst and Loucks [98] and Taylor [99], is that the most suitable criterion of performance is the efficiency of water use, θ:

$$\theta = \frac{\text{rate of photosynthesis}}{\text{rate of transpiration}}$$

There are, however, severe difficulties in applying this criterion. Most environments display seasonal and diurnal fluctuations of such magnitude that it would be necessary to postulate very demanding specifications for optimal leaf design: seasonally varying size and rapid movements during each day might need to be a feature for most temperate, mediterranean and subtropical species. Mountain plants, for example, must survive the winter without desiccation even though the soil water may be in the form of ice, and to achieve this, small leaves are required (large leaves develop higher leaf temperatures and so display a large leaf-to-air *VPD* and hence unacceptably high rates of water loss). In these winter conditions photosynthesis is unimportant as growth is nil, or small, and θ does not seem to be a useful parameter. In the spring when snow melts, water may be plentiful and conditions for photosynthesis favourable: then, water use seems practically irrelevant and only the rate of photosynthesis may be important for survival. For these and other reasons it seems unwise to apply θ except in a most general way, or in relation to plant breeding (see [107]).

In hot dry conditions with high radiation, as in the desert, a large leaf is likely to overheat (Fig. 4.1). This can to some extent be avoided if the stomatal resistance is very low so that a high rate of transpiration is achieved, but then the plant must support a very considerable root system. This would be unnecessary if the plant had small leaves which stayed close to the temperature of the air.

In arid conditions with low radiation, as in forest shade, small leaves would no longer have the advantage. On the contrary, dimly lit small leaves generally use more water than dimly lit large leaves (Fig. 4.2).

In cold conditions of high radiation a larger leaf should be at an advantage as its leaf temperature will be well above air temperature, facilitating high rates of photosynthesis and growth. On the other hand, this advantage would be offset by a much higher rate of water use, and so the specification of the optimum leaf involves a trade-off between the need to keep temperatures high and the need to conserve water. As already mentioned, the latter may be more important at certain times of year.

In cold conditions of low radiation, as in some northern forests, the advantage of high temperatures is no longer associated with large leaves, but large leaves, because of their high r_a, do have the advantage of relatively low rates of transpiration even when r_s is quite small (Fig. 4.2). It should not be assumed that water stress is unimportant in the ground flora: there is the complication of sunflecks, which cause a stepwise jump in the rate at which energy is absorbed. In one case, Woodward [102] has shown how this led to a dramatic increase in leaf temperature (much as would be expected from Fig. 4.1), associated with some degree of stomatal closure and a reduced rate of stem extension.

Perhaps the only clear prediction to be made from this sort of analysis is that smaller leaves are better adapted wherever temperature and radiation regimes are high and water is scarce. This prediction may be tested against recent data on leaf size, collected along the Orange River in South Africa [103]. Leaves were sampled in six vegetation types, all riverine forests, along a climatic gradient changing from temperature to hot and arid. One might have expected a gradual reduction in the size of leaves along this gradient. In fact, on progression to the drier zones, a diversification of leaf forms occurred, including both smaller and larger types. Trends were thus less clear than expected.

It is perhaps not surprising that trends in leaf size in mixed vegetation are not clear cut: individual species may differ in many other respects and occupy particular micro-environments, each with a unique energy regime. Within any one species, on the other hand, trends may be more strongly defined. Lewis [104] studied ecotypic variation in *Geranium sanguineum*: types from dry limestone had finely-dissected leaves, while woodland forms displayed larger leaves, composed of much broader units.

Richards [72] has drawn attention to the relatively uniform size distribution of leaves in the tropical rain forest, most of which are mesophyllous, with a trend towards microphylly as altitude increases. There are small but significant differences between the different strata, the lower strata containing larger leaves in general. As water is abundant in the rain forest and leaf water stress only mild, the use of water in transpiration seems once more to be unimportant. On the other hand, as a factor to be considered with respect to optimal leaf design, the direct evaporation from the leaf surface may be much more important, as the presence of water films on the leaf surface and infiltration into the mesophyll may both exert an adverse influence on gas exchange. There is very little published work on this point, and much more information on the photosynthetic attributes of rain forest species is required before an attempt can be made to explain the observed distribution of leaf size.

Another striking feature referred to by Richards [72] is that leaves from the tropical rain forest nearly always have entire margins. Observations in the Amazon by the botanical explorer Spruce, working at the turn of the century, include a note saying that the sight of a divided leaf was a 'rare treat' [72]. In the tree flora of the north temperate world, on the other hand, species with entire margins are rare, most individuals possessing serrated margins or divided leaves (Fig. 5.8). These features can perhaps be regarded as part of a trend towards microphylly and xeromorphy, essential if the plant is to avoid excessively large leaf-to-air vapour pressure differences, consequent high rates of transpiration and water stress.

5 cm

Fig. 5.8 Leaf form of trees native to the British Isles. All have serrated edges, except where the leaf is reduced to a needle.

This brief look at leaf size and shape suggests that the situation in the field may be more complex than the simple analysis of optimal leaf form can accommodate. Many other factors may be operating, and most environments are spatially and temporarily diverse, creating a general background 'noise' against which clear-cut relationships may be obscured.

Appendix 1 Specific heats

Specific heat of air $1010 \text{ J kg}^{-1} \text{ °C}^{-1}$
Specific heat of water vapour $1880 \text{ J kg}^{-1} \text{ °C}^{-1}$
Specific heat of carbon dioxide $850 \text{ J kg}^{-1} \text{ °C}^{-1}$

Appendix 2 Physical constants

Acceleration due to gravity 9.807 m s^{-2}
Stefan–Boltzmann constant $5.67 \times 10^{-8} \text{ W m}^{-2} \text{ K}^{-4}$
Avogadro's number $6.02204 \times 10^{23} \text{ mol}^{-1}$
Planck's constant $6.6261 \times 10^{-34} \text{ J s}$
Speed of light in vacuum $2.997925 \times 10^{8} \text{ m s}^{-1}$
Gas constant $8.3143 \text{ J mol}^{-1} \text{ K}^{-1}$

Appendix 3 Thermocouple data

Output of thermocouples made of the stated metals, with the reference junction kept at 0°C. Units: μV

	Temperature °C							
	−20	*−10*	*0*	*10*	*20*	*30*	*40*	*50*
Copper–constantan	−757	−383	0	391	789	1196	1611	2035
Chromel–constantan	−1151	−581	0	591	1192	1801	2419	3047
Chromel–alumel	−777	−392	0	397	798	1203	1611	2022
Iron–constantan	−995	−501	0	507	1019	1536	2058	2585

Appendix 4 Temperature dependent properties of air, water vapour and CO_2. (Source: Monteith [38])

°C	ρ_a ($kg\ m^{-3}$)	ρ_{as} ($kg\ m^{-3}$)	γ ($mbar\ C^{-1}$)	λ ($kJ\ kg^{-1}$)	κ ($mm^2\ s^{-1}$)	ν ($mm^2\ s^{-1}$)	D_v ($mm^2\ s^{-1}$)	D_c ($mm^2\ s^{-1}$)
−5	1.316	1.314	0.643	2513	18.3	12.9	20.5	12.4
0	1.292	1.289	0.646	2501	18.9	13.3	21.2	12.9
5	1.269	1.265	0.649	2489	19.5	13.7	22.0	13.3
10	1.246	1.240	0.652	2477	20.2	14.2	22.7	13.8
15	1.225	1.217	0.655	2465	20.8	14.6	23.4	14.2
20	1.204	1.194	0.658	2454	21.5	15.1	24.2	14.7
25	1.183	1.169	0.662	2442	22.2	15.5	24.9	15.1
30	1.164	1.145	0.665	2430	22.8	16.0	25.7	15.6
35	1.146	1.121	0.668	2418	23.5	16.4	26.4	16.0
40	1.128	1.096	0.671	2406	24.2	16.9	27.2	16.5
45	1.110	1.068	0.675	2394	24.9	17.4	28.0	17.0

ρ_a Density of dry air
ρ_{as} Density of water-saturated air
λ Latent heat of vaporization
γ The psychrometer constant

κ Thermal diffusivity of dry air
ν Kinematic viscosity of dry air
D_v Diffusion coefficient of water vapour in air
D_c Diffusion coefficient of CO_2 in air

Appendix 5 Saturated vapour pressure (e_s) and black body radiation (σT^4)

°C	e_s (mbar) water	ice	σT^4 (W m^{-2})	°C	e_s (mbar) water	σT^4 (W m^{-2})
−10	2.86	2.60	272.1	20	23.37	419.0
−9	3.11	2.84	276.3	21	24.86	424.8
−8	3.35	3.10	280.5	22	26.43	430.6
−7	3.62	3.39	284.7	23	28.08	436.4
−6	3.91	3.68	289.0	24	29.83	442.4
−5	4.21	4.01	293.4	25	31.67	448.3
−4	4.54	4.37	297.8	26	33.61	454.4
−3	4.90	4.76	302.2	27	35.65	460.5
−2	5.27	5.17	306.7	28	37.79	466.7
−1	5.68	5.62	311.3	29	40.05	472.9
0	6.11	6.11	315.9	30	42.43	479.2
1	6.57		320.5	31	44.93	485.5
2	7.05		325.2	32	47.55	492.0
3	7.57		330.0	33	50.30	498.4
4	8.13		334.8	34	53.20	505.0
5	8.72		339.6	35	56.23	511.6
6	9.35		344.5	36	59.42	518.3
7	10.01		349.5	37	62.76	525.0
8	10.72		354.5	38	66.26	531.8
9	11.47		359.6	39	69.93	538.7
10	12.27		364.7	40	73.78	545.6
11	13.11		369.9	41	77.80	552.6
12	14.02		375.1	42	82.01	559.7
13	14.97		380.4	43	86.42	566.8
14	15.98		385.8	44	91.03	574.0
15	17.04		391.2	45	95.85	581.3
16	18.17		396.6	46	100.89	588.6
17	19.37		402.1	47	106.16	596.0
18	20.63		407.7	48	111.65	603.5
19	21.96		413.3	49	117.40	611.1
				50	123.40	618.7
				51	129.65	626.4
				52	136.17	634.1
				53	142.98	642.0
				54	150.07	649.9
				55	157.46	657.9

Appendix 6 Useful formulae

1. A good approximation for calculating the saturated vapour pressure e_s (mbar) at any temperature between -5 and $50°C$:

$$e_s = \exp[a + (bT - c)/(T - d)]$$

where T is the temperature (K), and the empirical coefficients are

$a = 1.80956664$, $b = 17.2693882$, $c = 4717.306081$, $d = 35.86$.

2. Relationship between vapour pressure and aspirated wet bulb temperature at an atmospheric pressure of 1000 mbar:

$$e = e_s - 0.666(T_{dry} - T_{wet})$$

where T_{dry} and T_{wet} are the dry and wet bulb temperatures respectively ($°C$).

3. To obtain absolute water content χ (g m^{-3}) from e_s (mbar):

$$\chi = 217\, e/T$$

where T is the temperature (K).

4. Ratios of boundary layer resistances:

$$r_a^{H_2O}/r_a^H = (\kappa/D_{H_2O})^{\frac{2}{3}} = 0.93$$
$$r_a^{CO_2}/r_a^H = (\kappa/D_{CO_2})^{\frac{2}{3}} = 1.32$$

5. To calculate the sun's angle at any location and time:

For any locality and time the sun's angular elevation a and its angular distance from the geographical south b may be found from the equations:

$$\sin a = \sin \varphi \sin \delta + \cos \varphi \cos \delta \cos h$$
$$\sin b = \cos \delta \sin h / \cos a$$

where h is the hour angle of the sun, being the difference between the given time and the time of apparent noon, expressed in degrees where 1 h is $15°$, φ is the latitude and δ is the sun's angle of declination for that time of year. Approximate values of δ can be found by interpolating from the following table, but more accurate information is published in almanacs.

Solar declination on the first day of each month (degrees)

Jan.	-23.1	Apr.	$+4.1$	July	$+23.2$	Oct.	$+2.8$
Feb.	-17.3	May	$+14.8$	Aug.	$+18.3$	Nov.	-14.1
Mar.	-8.0	June	$+21.9$	Sept.	$+8.6$	Dec.	-21.6

Appendix 7   Symbols and abbreviations

All symbols for quantities are set in *italics*, except for those which denote flux densities – these are in **bold** type. Roman type is used for mathematical operators, units, and the dimensionless numbers like Re which would otherwise look like two symbols multiplied together.

Roman alphabet

a	coefficient of expansion ($°C^{-1}$)
C	convective heat flux ($W\ m^{-2}$)
c	carbon dioxide concentration ($cm^3\ m^{-3}$)
c_p	specific heat at constant pressure ($1010\ J\ kg^{-1}\ °C^{-1}$)
D	diffusion coefficient ($mm^2\ s^{-1}$)
d	characteristic dimension (mm)
e	vapour pressure of water vapour in air (mbar or kPa)
$e_s(T)$	saturation vapour pressure of water vapour at temperature T (mbar or kPa)
E	evaporation rate ($kg\ m^{-2}\ s^{-1}$)
F	flux density of mass ($kg\ m^{-2}\ s^{-1}$) or radiation ($W\ m^{-2}$)
f	frequency (s^{-1})
G	heat flux by conduction ($W\ m^{-2}$)
Gr	Grashof number (no dimensions)
G	conductance ($m\ s^{-1}$).
g	acceleration caused by gravity ($9.81\ m\ s^{-2}$)
h	Planck's constant ($6.63 \times 10^{-34}\ J\ s$)
h	height of vegetation
K	turbulent transfer coefficient ($m^2\ s^{-1}$) with subscripts to indicate water (E), heat (H) and mass (m)
k	von Karman's constant (0.41)
LAI	leaf area index (leaf area per land area)
l	length (m or mm)
PAR	photosynthetically active radiation, usually taken as that within the waveband 400–750 nm ($W\ m^{-2}$ or $mol\ m^{-2}\ s^{-1}$)
P	rate of photosynthesis ($W\ m^{-2}$ if in energy terms, otherwise $mg\ CO_2\ m^{-2}\ s^{-1}$)
R	resistance of a chain or network of resistors ($s\ m^{-1}$)
R	net flux density of all-wave radiation ($W\ m^{-2}$) as measured with a net radiometer. Also the net ~~energy~~ absorbed by a leaf or a canopy of leaves ($W\ m^{-2}$)
r_a	boundary layer or aerodynamic resistance to transfer of water, heat, mass or momentum ($s\ m^{-1}$)
r_b	excess resistance ($s\ m^{-1}$)
r_s	stomatal resistance ($s\ m^{-1}$)
r_{st}	stomatal resistance estimated for a leaf canopy ($s\ m^{-1}$)
Re	Reynolds number (no dimensions)
Ri	Richardson number (no dimensions)

S	rate of transfer of heat to store (W m^{-2})
s	rate of change of saturation vapour pressure with temperature (mbar °C^{-1})
T	temperature
T_s	surface temperature
u	wind speed (m s^{-1})
u_*	friction velocity (m s^{-1})
VPD	water vapour pressure difference between evaporating surface and the air above. Also written $e_s(T) - e$ (mbar or kPa)
z	height above the ground (m)
z_0	roughness length (m)

Greek alphabet

γ (gamma)	psychrometer constant (0.66 mbar °C^{-1})
ε (epsilon)	emissivity (no dimensions – a fraction)
ζ (zeta)	measure of red:far-red ratio
κ (kappa)	thermal diffusivity of air (mm^2 s^{-1}, see Appendix 4)
λ (lambda)	latent heat of vaporization of water (J g^{-1}, see Appendix 4)
ν (nu)	coefficient of kinematic viscosity of air (mm^2 s^{-1}, see Appendix 4)
ρ (rho)	density of a gas (kg m^{-3}, see Appendix 4)
σ (sigma, small)	Stefan–Boltzmann constant (5.67×10^{-8} W m^{-2} K^{-4})
Σ (sigma, large)	sum of a series
χ (chi)	concentration (kg m^{-3})

Appendix 8 Units

bar	unit of pressure, not SI but still widely used (1 bar = 10^5 Pascals)
°C	degrees Celsius, also called Centigrade
E	the Einstein, a unit of photoenergy depending on the frequency of the radiation: Avogadro's number of quanta (see Section 1.1.4)
g	gram
K	degrees Kelvin (°C = K − 273.15)
m	metre
mol	the mole, strictly a unit of quantity related to molecular mass but also used for Avogadro's number of quanta
N	the Newton, a unit of force
Pa	the Pascal, a force of one Newton distributed over one square metre
s	second
W	the Watt, a unit of power (1 J s^{-1})
yr	year

Appendix 9 Metric multiples and submultiples

T	tera-	10^{12}		d	deci-	10^{-1}
G	giga-	10^{9}		c	centi-	10^{-2}
M	mega-	10^{6}		m	milli-	10^{-3}
my	myria-	10^{4}		μ	micro-	10^{-6}
k	kilo-	10^{3}		n	nano-	10^{-9}
h	hecto-	10^{2}		p	pico-	10^{-12}
da	deca-	10				

It is usually recommended that the units of the SI (International System of units) should be modified by prefixes that represent steps of 10^{3} and 10^{-3} only. However, cm is still widely used.

References

[1] Gates, D.M. (1980), *Biophysical Ecology*, Springer–Verlag, New York.

[2] List, R.J. (1963), *Smithsonian Meteorological Tables*, Smithsonian Institution, Washington.

[3] Ripley, E.A. and Redman, R.E. (1976), Grassland, in *Vegetation and the Atmosphere*, 2: *Case studies* (Monteith, J.L., ed.) Academic Press, London, pp. 349–398.

[4] Dresner, S. (1971), *Units of Measurement*, Harvey Miller and Medcalf, Aylesbury.

[5] Incoll, L.D., Long, S.P. and Ashmore, M.R. (1977), SI units in publications in plant science: commentaries in Plant Science **28**. *Current Advances in Plant Science*, **3**, 331–343.

[6] Keeling, C.D., Bacastwo, R.B., Bainbridge, A.E. *et al.* (1976), Atmospheric carbon dioxide variations at Mauna Loa Observatory, Hawaii. *Tellus*, **28**, 538–551.

• [7] Almquist, E. (1974), An analysis of global pollution. *Ambio*, **3**, 161–167.

[8] Liu, S.C., Cicerone, R.J., Donahue, T.M. and Chameides, W.L. (1977), Sources and sinks of atmospheric N_2O and the possible ozone reduction due to industrial fixed nitrogen fertilizers. *Tellus*, **29**, 251–263.

[9] Lovelock, J.E. and Margulis, L. (1973), Atmospheric homeostasis by and for the biosphere: the gaia hypothesis. *Tellus*, **26**, 2–10.

[10] Lovelock, J.E. (1979), *Gaia: a New Look at Life on Earth*, Oxford University Press, Oxford.

[11] Körner, Ch. and Mayr, R. (1981), Stomatal behaviour in alpine plant communities between 600 and 2600 metres above sea level. *Plants and their Environment, 21st Symposium of the British Ecological Society* (Grace, J., Ford, E.D. and Jarvis, P.G., eds), pp. 205–218, Blackwell Scientific Publications, Oxford.

[12] Cohen, S.S., Gale, J., Shmida, A. *et al.* (1981), Xeromorphism and potential rate of transpiration on Mount Hermon, an east Mediterranean mountain. *Journal of Ecology*, **69**, 391–403.

[13] Tranquillini, W. (1979), *Physiological Ecology of the Alpine Timberline; tree existence at high altitudes with special reference to the European Alps*, Springer–Verlag, New York.

[14] Harding, R.J. (1979), Radiation in the British Uplands. *Journal of Applied Ecology*, **16**, 161–170.

[15] Mooney, H.A., Strain, E.R. and West, M. (1966), Photosynthetic efficiency at reduced carbon dioxide tensions. *Ecology*, **47**, 490 491.

[16] Grace, J. (1977), *Plant Response to Wind*, Academic Press, London.

[17] Stanhill, G. (1981), The size and significance of differences in the radiation balance of plants and plant communities. In *Plants and their Atmospheric Environment, 21st Symposium of the British Ecological Society* (Grace, J., Ford, E.D. and Jarvis, P.G., eds), pp 57–74, Blackwell Scientific Publications, Oxford.

[18] Thomas, D.A. and Barber, H.N. (1974), Studies on leaf characteristics of

a cline of *Eucalyptus urnigera* from Mount Wellington, Tasmania. 2: Reflection, transmission and the absorption of radiation. *Australian Journal of Botany*, **22**, 701–707.

[19] Duncan, W.G., Loomis, R.S., Williams, W.A. and Hanau, R. (1967), A model for simulating photosynthesis in plant communities. *Hilgardia*, **38**, 181–205.

[20] Monteith, J.L. (1965), Light distribution and photosynthesis in field crops. *Annals of Botany*, **29**, 17–37.

[21] de Wit, C.T. (1965), Photosynthesis of leaf canopies. *Agricultural Research Report No. 663*, Centre for Agricultural Publications and Documentation, Wageningen.

[22] Cowan, I.R. (1968), The interception and absorption of radiation in plant stands. *Journal of Applied Ecology*, **5**, 367–379.

[23] Norman, J.M. and Jarvis, P.G. (1975), Photosynthesis in Sitka spruce (*Picea sitchensis* (Bong) Carr.). 5: Radiation penetration theory and a test case. *Journal of Applied Ecology*, **12**, 839–878.

[24] Norman, J.M. (1979), Modelling the complete crop canopy. In *Modification of the Aerial Environment* (Barfield, B.J. and Gerber, J.F., eds), pp. 242–277, American Society of Agricultural Engineers Monograph 2, St. Joseph, Michigan.

[25] Lang, A.R.G. and Begg, J.E. (1979), Movements of *Helianthus annuus* leaves and heads. *Journal of Applied Ecology*, **16**, 299–306.

[26] Smith, H. (1981), Light quality as an ecological factor. In *Plants and their Atmospheric Environment, 21st Symposium of the British Ecological Society* (Grace, J., Ford, E.D. and Jarvis, P.G., eds), pp. 93–110, Blackwell Scientific Publications, Oxford.

[27] Smith, H. (1975), *Phytochrome and Photomorphogenesis*, McGraw-Hill, London.

[28] Leech, R. (1976), The photosynthetic apparatus of higher plants. *Plant Structure, Function and Adaptation* (Hall, M.A., ed.), pp. 125–156, Macmillan, London and Basingstoke.

[29] Boardman, N.K. (1977), Chloroplasts – structure and photosynthesis. *The Molecular Biology of Plant Cells* (Smith, H., ed.), pp. 85–104, Blackwell Scientific Publications, Oxford.

[30] Björkman, O. (1981), The response of photosynthesis to temperature. *Plants and their Atmospheric Environment, 21st Symposium of the British Ecological Society* (Grace, J., Ford, E.D. and Jarvis, P.G., eds), pp. 273–301, Blackwell Scientific Publications, Oxford.

[31] Long, S.P., Incoll, L.D. and Woolhouse, H.W. (1975), C_4 photosynthesis in plants from cool temperate regions, with particular reference to *Spartina townsendii*. *Nature*, **257**, 622–624.

[32] Dunn, R., Long, S.P. and Thomas, S.M. (1981), The effect of temperature on the growth and photosynthesis of the temperate C_4 grass *Spartina townsendii*. *Plants and their Atmospheric Environment, 21st Symposium of the British Ecological Society* (Grace, J., Ford, E.D. and Jarvis, P.G., eds), pp. 303–311. Blackwell Scientific Publications, Oxford.

[33] Biscoe, P.V. and Gallagher, J.N. (1977), Weather, dry matter production and yield. *Environmental Effects on Crop Physiology* (Landsberg, J.J. and Cutting, C.V., eds), pp. 75–100, Academic Press, London.

[34] Monteith, J.L. (1981), Coupling of plants to the atmosphere. *Plants and their Atmospheric Environment, 21st Symposium of the British*

Ecological Society (Grace, J., Ford, E.D. and Jarvis, P.G., eds), pp. 1–29, Blackwell Scientific Publications, Oxford.

[35] Grace, J. and Wilson, J. (1976). The boundary layer over a *Populus* leaf. *Journal of Experimental Biology*, **27**, 231–241.

[36] Jarvis, P.G. (1971), The estimation of resistances to carbon dioxide transfer. *Plant Photosynthetic Production: Manual of Methods* (Sestak, Z., Catsky, J. and Jarvis, P.G., eds), pp. 566–631, Junk, The Hague.

[37] Grace, J., Fasehun, F.E. and Dixon, M.A. (1980), Boundary layer conductance of the leaves of some tropical timber trees. *Plant, Cell and Environment*, **3**, 443–450.

[38] Monteith, J.L. (1973), *Principles of Environmental Physics*, Edward Arnold, London.

[39] Biscoe, P.V., Clark, J.A. Gregson, K. *et al.* (1975), Barley and its environment. 1: Theory and practice. *Journal of Applied Ecology*, **12**, 227–257.

[40] Thom, A.S. (1975), Momentum, mass and heat exchange in plant communities. *Vegetation and the Atmosphere, Vol. 1: Principles* (Monteith, J.L., ed.), pp. 57–109, Academic Press, London.

[41] Garrett, J.R. (1977), Aerodynamic Roughness and Mean Monthly Surface Stress over Australia, *Division of Atmospheric Physics Technical Paper*, CSIRO Australia, **29**, 1–19.

[42] Paulson, C.A. (1970), The mathematical representation of wind speed and temperature profiles in the unstable atmosphere surface layer. *Journal of Applied Meteorology*, **9**, 857–861.

[43] Dyer, A.J. (1974), A review of flux-profile relationships. *Boundary Layer Meteorology*, **7**, 363–372.

[44] Businger, J.A. (1975), Aerodynamics of vegetated surfaces. *Heat and Mass Transfer in the Biosphere* (de Vries, D.A. and Afgan, N.H., eds), pp. 139–165, Wiley, New York.

[45] Yaglom, A.M. (1977), Comments on wind and temperature flux profile relationships. *Boundary Layer Meteorology*, **11**, 89–102.

[46] Landsberg, J.J. and James, G.B. (1971), Wind profiles in plant canopies: studies on an analytical model. *Journal of Applied Ecology*, **8**, 729–742.

[47] Legg, B.J. (1975), Turbulent diffusion within a wheat canopy. 1: Measurement using nitrous oxide. *Quarterly Journal of the Royal Meteorological Society*, **101**, 597–610.

[48] Druilhet, A., Perrier, A., Fontan, J. and Laurent, J.L. (1971), Analysis of turbulent transfers in vegetation: use of thoron for measuring the diffusivity profiles. *Boundary Layer Meteorology*, **2**, 173–187.

[49] Jeffree, C.E., Baker, E.A. and Holloway, P.J. (1976), Origins of the Fine Structure of Plant Epicuticular Waxes. *Microbiology of Aerial Plant Surfaces* (Dickinson, C.H. and Preece, T.F., eds), pp. 119–158, Academic Press, London.

[50] Grace, J. (1974), The effect of wind on grasses. 1: Cuticular and stomatal transpiration. *Journal of Experimental Botany*, **25**, 542–551.

[51] Thompson, J.R. (1974), The effect of wind on grasses. 2: Mechanical damage in *Festuca arundinacea* Schreb. *Journal of Experimental Botany*, **25**, 965–972.

[52] Rutter, A.J. (1975), The hydrological cycle in vegetation. *Vegetation and the Atmosphere, Vol. 1: Principles* (Monteith, J.L., ed.), pp. 111–154. Academic Press, London.

[53] Proctor, M.C.F. (1981), Diffusion resistances in Bryophytes. *Plants and their Atmospheric Environment, 21st Symposium of the British Ecological Society* (Grace, J., Ford, E.D. and Jarvis, P.G., eds), pp. 219–235, Blackwell Scientific Publications, Oxford.

[54] Jones, H.G. and Norton, T.A. (1981), The role of internal factors in controlling evaporation from intertidal algae. *Plants and their Atmospheric Environment, 21st Symposium of the British Ecological Society* (Grace, J., Ford, E.D. and Jarvis, P.G., eds), pp. 231–235, Blackwell Scientific Publications, Oxford.

[55] Körner, Ch., Scheel, J.A. and Bauer, H. (1979), Maximum leaf diffusive conductance in vascular plants. *Photosynthetica*, **13**, 45–82.

[56] Meidner, H. and Mansfield, T.A. (1968), *Physiology of Stomata*, McGraw-Hill, London.

[57] Jarvis, P.G. and Morison, J.I.L. (1981), The control of transpiration and photosynthesis by the stomata. *Stomatal Physiology Society for Experimental Biology Seminar, Vol. 8* (Jarvis, P.G. and Mansfield, T.A., eds), pp. 247–277, Cambridge University Press, London.

[58] Stiles, W. (1970), A diffusive resistance porometer for field use. *Journal of Applied Ecology*, **7**, 617–622.

[59] Jarvis, P.G. (1981), Stomatal conductance, gaseous exchange and transpiration. *Plants and their Atmospheric Environment, 21st Symposium of the British Ecological Society* (Grace, J., Ford, E.D. and Jarvis, P.G.), pp. 175–204, Blackwell Scientific Publications, Oxford.

[60] Grace, J., Okali, D.U.U. and Fasehun, F.E. (1982), Stomatal conductance of two tropical trees during the wet season in Nigeria. *Journal of Applied Ecology*, **19**, 659–670.

[61] Jarvis, P.G. (1976), The interpretation of the variations in leaf water potential and stomatal conductance found in canopies in the field. *Philosophical Transactions of the Royal Society, London, B*, **273**, 593–610.

[62] Gaastra, P. (1959), Photosynthesis of crop plants as influenced by light, carbon dioxide, temperature and stomatal diffusion resistance. *Mededelingen van de Landbouwhogeschooll te Wageningen, Nederland*, **59**, 1–68.

[63] Stewart, J.B. and Thom, A.S. (1973), Energy budgets in pine forest. *Quarterly Journal of the Royal Meteorological Society*, **99**, 154–170.

[64] Monteith, J.L. (1981), Evaporation and surface temperature. *Quarterly Journal of the Royal Meteorological Society*, **107**, 1–27.

[65] Yamaoka, Y. (1958), Total transpiration from a forest. *Transactions of the American Geophysical Union*, **39**, 266–272.

[66] Monteith, J.L. (1965), Evaporation and Environment. *The State and Movement of Water in Living Organisms, 19th Symposium of the Society for Experimental Biology* (Fogg, G.E., ed.), pp. 205–234, Cambridge University Press, London.

[67] Waring, R.H., Whitehead, D. and Jarvis, P.G. (1980), Comparison of an isotopic method and the Penman–Monteith equation for estimating transpiration from Scots pine. *Canadian Journal of Forest Research*, **10**, 555–558.

[68] Cernusca, A. and Seeber, M.C. (1981), Canopy structure, microclimate and the energy budget in different alpine communities. *Plants and their Atmospheric Environment, 21st Symposium of the British Ecological*

Society (Grace, J., Ford, E.D. and Jarvis, P.G., eds), pp. 75–81, Blackwell Scientific Publications, Oxford.

[69] Whitehead, D., Okali, D.U.U. and Fasehun, F.E. (1981), Stomatal response to environmental variables in two tropical forest species during the dry season in Nigeria. *Journal of Applied Ecology*, **18**, 571–588.

[70] Grace, J., Malcolm, D.C. and Bradbury, I.K. (1975), The effect of wind and humidity on leaf diffusive resistance in Sitka spruce seedlings. *Journal of Applied Ecology*, **12**, 931–940.

[71] Russell, G. (1980), Crop evaporation, surface resistance and soil water status. *Agricultural Meteorology*, **21**, 213–226.

[72] Richards, P.W. (1952), *The Tropical Rain Forest*, Cambridge University Press, London.

[73] Raunkiaer, C. (1909), Formationsundersogelse og Formationsstatisk. *Botanisk Tidsskrift*, **30**, 20–132.

[74] Raunkiaer, C. (1934), *The Life Forms of Plants and Statistical Plant Geography*, University Press, Oxford.

[75] Gimingham, C.H. (1951), The use of life form and growth form in the analysis of community structure, as illustrated by a comparison of two dune communities. *Journal of Ecology*, **39**, 396–406.

[76] Whitehead, F.H. (1959), Vegetational change in response to alterations of surface roughness on Monte Maiella, Italy. *Journal of Ecology*, **47**, 603–606.

[77] Lambert, J.M. (1972), Theoretical models for large-scale vegetation survey. *Mathematical Models in Ecology, 12th Symposium of the British Ecological Society* (Jeffers, J.N.R., ed.), pp. 87–109, Blackwell Scientific Publications, Oxford.

[78] Richards, P.W., Tansley, A.G. and Watt, A.S. (1939), *The recording of structure, life form and flora of tropical forest communities as a basis for their classification*, Imperial Forestry Institute, Paper 19, University of Oxford.

[79] Bailey, I.W. and Sinnot, E.W. (1916), The climatic distribution of certain types of Angiosperm leaves. *American Journal of Botany*, **3**, 24–39.

[80] Clausen, J., Keck, D.D. and Hiesey, W.M. (1948), *Experimental studies on the nature of species. 3: Environmental responses of climatic races of Achillea*. Carnegie Institute, Washington. Publication no. 581.

[81] Grant, S.A. and Hunter, R.F. (1962), Ecotypic differentiation of *Calluna vulgaris* (L) in relation to altitude. *New Phytologist*, **61**, 44–55.

[82] Aston, J.L. and Bradshaw, A.D. (1966), Evolution in closely adjacent plant populations. 2: *Agrostis tenuis* in maritime habitats. *Heredity*, **21**, 649–664.

[83] Salisbury, F.B. and Spomer, G.G. (1964), Leaf temperatures of alpine plants in the field. *Planta*, **60**, 497–505.

[84] Clausen, J. (1965), Population studies of alpine and subalpine races of conifers and willows in the California High Sierra Nevada. *Evolution*, **19**, 56–68.

[85] Wardle, P. (1968), Engleman spruce (*Picea engelmanii* Engel.) at its upper limits on the front range, Colorado. *Ecology*, **49**, 483–495(F).

[86] Jarvis, P.G., James, G.B. and Landsberg, J.J. (1976), Coniferous forest. *Vegetation and the Atmosphere, Vol. 2, Case Studies* (Monteith, J.L., ed.), pp. 171–240. Academic Press, London.

[87] Watts, W.R. (1972), Leaf extension on *Zea mays*. 2: Leaf extension in response to independent variation of the temperature of the apical meristem, of the air around the leaves, and of the root zone. *Journal of Experimental Botany*, **76**, 713–721.

[88] Peacock, J.M. (1975), Temperature and leaf growth in *Lolium perenne*. 2: The site of temperature perception. *Journal of Applied Ecology*, **12**, 115–124.

[89] Oura, H. (1953), On the capture of fog particles by a forest. *Studies on Fogs in Relation to Fog-preventing Forests* (Hori, T., ed.), pp. 253–259, Tanne Trading Company, Sapporo, Japan.

[90] Boyce, S.G. (1954), The salt spray community. *Ecological Monographs*, **24**, 29–67.

[91] Levitt, J. (1972), *Responses of Plants to Environmental Stresses*, Academic Press, New York.

[92] Lange, O.L. (1957), Hitzeresistenz und Blattemperaturen mauretanischer Wüstenpflanzen. *Berichte der Deutschen Botanischen Gesellschaft*, **70**, 31–32.

[93] Lange, O.L. (1961), Die Hitzeresistenz einheimischer immer-und wintergrüner Pflanzen im Jahreslauf. *Planta*, **56**, 666–683.

[94] Lange, O.L. (1959), Untersuchungen über Wärmehaushalt und Hitzeresistenz mauretanischer Wüsten-und Savannenpflanzen. *Flora* (Jena), **147**, 595–651.

[95] Larcher, W. (1980), *Physiological Plant Ecology*, 2nd edn, Springer–Verlag, Amsterdam.

[96] Stoutjesdijk, P. (1970), Some measurements of leaf temperatures of tropical and temperate plants and their interpretation. *Acta Botanica Neerlandica*, **19**, 373–384.

[97] Lange, O.L., Schulze, E.D., Evanari, M., Kappen, L. and Buschboom, U. (1978), The temperature-related photosynthetic capacity of plants under desert conditions. 3: Ecological significance of the seasonal changes of the photosynthetic response to temperature. *Oecologia*, **34**, 89–100.

[98] Parkhurst, D.F. and Loucks, D.L. (1972), Optimal leaf size in relation to environment. *Journal of Ecology*, **60**, 505–537.

[99] Taylor, S.E. (1975), Optimal leaf form. In *Perspectives of Biophysical Ecology* (Gates, D.M., ed.), pp. 73–86, Springer–Verlag, New York.

[100] Campbell, G.S. (1977), *An Introduction to Environmental Biophysics*. Springer–Verlag, New York.

[101] Jones, H.G. (1976), Crop characteristics and the ratio between assimilation and transpiration. *Journal of Applied Ecology*, **13**, 605–622.

[102] Woodward, F.I. (1981), Shoot extension and water relations of *Circaea lutetiana* in sunflecks. In *Plants and their Atmospheric Environment, 21st Symposium of the British Ecological Society* (Grace, J., Ford, E.D. and Jarvis, P.G., eds), pp. 83–91, Blackwell Scientific Publications, Oxford.

[103] Werger, M.J.A. and Ellenbroek, G.A. (1978), Leaf size and leaf consistence of a Riverine Forest formation along a climatic gradient. *Oecologia*, **34**, 297–308.

[104] Lewis, M.C. (1972), The physiological significance of variation in leaf structure. *Science Progress*, **60**, 25–51.

[105] Fritschen, L.J. and Lloyd, W.G. (1979), *Environmental Instrumentation*, Springer–Verlag, New York.

Index